T0201655

Playfair's Commercial and Political Atlas
and Statistical Breviary

A scientific revolution began at the end of the 18th century with the invention and popularization of the graphic display of data by the remarkable Scot William Playfair. His marvelous *Atlas* showed how much could be learned if one plotted data atheoretically and looked for suggestive patterns. Those patterns provide evidence, albeit circumstantial, on which to build new science. Playfair's work has much to teach us, but finding a copy of it is very difficult. This full-color reproduction of two of his classic works, with new explanatory material, makes Playfair's wisdom widely available for the first time in two centuries.

Howard Wainer is Distinguished Research Scientist for the National Board of Medical Examiners and Adjunct Professor of Statistics at the Wharton School of the University of Pennsylvania. He is the author of 15 previous books, most recently, *Graphic Discovery: A Trout in the Milk and Other Visual Adventures* (2005).

Ian Spence is Professor of Psychology at the University of Toronto. His current research includes experimentation in human–computer interaction and visual memory. He is the author of more than 100 research articles and monographs.

"William Playfair's *Commercial and Political Atlas* and his *Statistical Breviary* are among the most important works in the entire history of statistical graphics and data visualization. Here we find the origin of the modern graphical forms most widely used today – the pie chart, line graph and bar chart – and Playfair used these with great skill to make his (largely economic) data 'speak to the eyes.' While some of his graphs have been reprinted, often badly, in historical studies, few people have been able to study the very few extant complete copies of these works to see the scope (and beauty) of Playfair's graphical innovations together, and in original context. At least as important, a modern reader will want to read Playfair's words to see how he faced the challenge of presenting his novel charts to his audience around 1800.

"Spence and Wainer have done a great service to all those interested in visual information display and its history, first by providing high-resolution copies of Playfair's main works, and second by giving us a window on 'what he was thinking.'"
– Michael Friendly, York University

"William Playfair was the great pioneer of statistical graphics. The striking inventions of these path-breaking books show already in 1801 the medium's capacity to distill large amounts of data into easily-grasped pictures, as well as the dangers of graphical misinformation. If Playfair's time series figures misleadingly portray Britain's economic condition through emphasis upon balance of trade, and his pie charts risk deceiving through submersion of magnitudes in service of proportions, all of these nonetheless convey the author's polemical message accurately and brilliantly. Playfair's text brings its fascinating author and his time to life, and his beautiful colored charts are functional artistic creations of a high order. These books are the well-spring of modern graphical display, warts and all."
– Stephen Stigler, University of Chicago

The Commercial and Political Atlas and Statistical Breviary

William Playfair

Edited and Introduced by

Howard Wainer   Ian Spence
National Board of   *University of Toronto*
Medical Examiners

CAMBRIDGE
UNIVERSITY PRESS

CAMBRIDGE UNIVERSITY PRESS

Cambridge, New York, Melbourne, Madrid, Cape Town, Singapore, São Paulo

Cambridge University Press
40 West 20th Street, New York, NY 10011-4211, USA

www.cambridge.org
Information on this title: www.cambridge.org/9780521855549

First published 2005

Printed in Hong Kong by Golden Cup

A catalog record for this publication is available from the British Library.

Library of Congress Cataloging-in-Publication Data

Playfair, William, 1759–1823.
Playfair's Commercial and political atlas and Statistical breviary / William Playfair; edited and
introduced by Howard Wainer and Ian Spence.
p. cm.
A full color reproduction of the Atlas, 3rd ed., and of the Breviary, both published in
London by J. Wallis, 1801.
"Bibliography: variations of the Atlas and Breviary": p. .
Includes bibliographical references.
ISBN 0-521-85554-3
1. Great Britain – Statistics – Early works to 1800. 2. Great Britain – Commerce – Early works to
1800. 3. Great Britain – Commerce – History – 18th century. 4. Europe – Statistics – Early
works to 1800. 5. Statistics – Graphic methods. I. Title: Commercial and political atlas and
Statistical breviary. II. Title: Playfair's Statistical breviary. III. Title: Statistical breviary.
IV. Playfair, William, 1759–1823. Statistical breviary. V. Wainer, Howard. VI. Spence, Ian,
1944– VII. Playfair, William, 1759–1823. Statistical breviary. VIII. Title.
HF3501.5.P58 2005
314′.09′033–dc22   2005046971

ISBN-13   978-0-521-85554-9 hardback
ISBN-10   0-521-85554-3 hardback

To
Linda Steinberg
and
Elsa Marziali

Sine quibus non

Preface

To those interested in the effective visual communication of quantitative phenomena, William Playfair's *Atlas* is like The Bible: an ancient and revered book that is often cited but rarely read. The *Atlas* has been difficult to find and read simply because it is old and rare – copies that still exist are either prized by private collectors or stored safely within the confines of rare book libraries. Happily, after more than two centuries, the *Atlas* is now available for anyone who would like one. While we are pleased to have played our part in this rebirth, our role was not that of a Gutenberg (that honor belongs to Cambridge University Press). We play the role of a pair of graphical Gideons; the reproduction of this masterwork at an affordable price means that it will now be widely available to spread the gospel according to William Playfair.

The volume you hold in your hand contains more than Playfair's *Atlas*, although that alone would have been enough to cherish. It also includes his *Statistical Breviary*. During the span of sixteen years, from 1786 to 1801, Playfair revised and republished his *Atlas* twice. We have chosen to reproduce here the third and most mature edition. It was published in the same year as his *Statistical Breviary*, and so putting them together provides a fuller picture of his thinking at that time. As Playfair himself said, "the two go with great Propriety together." In addition to the facsimiles, we have prepared an introduction relating many of the fascinating, and often surprising, details of the life led by William Playfair. The introduction also illuminates the contents of these volumes, discussing the technology behind their printing and highlighting some of Playfair's conceptual breakthroughs, as well as some of the difficulties, idiosyncrasies, and infelicities.

Bringing this republication to fruition required the assistance of many, and it is our pleasure to take this opportunity to express our most profound gratitude for their efforts.

This reproduction is copied from the *Atlas* and *Breviary* that are owned by the University of Pennsylvania, and we therefore first would like to thank the Annenberg Rare Book and Manuscript Library of the University of Pennsylvania. Among those at the library who warrant special mention are Greg Bear and his staff at the Schoenberg Center for Electronic Text and Image. *Primus inter pares* is John Pollack; he was generous with both his time and energy in his efforts to provide the finest possible copy.

Second in line for thanks is Abba Krieger, Chairman of the Department of Statistics at the University of Pennsylvania. Abba's support for a graduate seminar on statistical graphics (Statistics 991) was the catalyst that set us on a path leading initially to the rare books library and ultimately to scanning the works of interest for broader use.

The W. K. Kellogg Foundation, specifically Anne Petersen, Senior Vice President for Programs, deserves special thanks for providing the financial subsidy that allowed the publisher not only to assume the risk of reproducing a two-hundred-year-old book requiring high production costs, but to do so at a sale price that makes it accessible to a broad range of readers.

Our gratitude also to the National Board of Medical Examiners in general and to Donald Melnick, President, and Ronald Nungester, Senior Vice President, in particular, for support of and enthusiasm for the project despite its only very tenuous connection to the goals of the Board, and to Melissa Margolis for her fine editorial eye.

We thank the University of Toronto for its support in the form of research leaves to Ian Spence to pursue his biographical work on Playfair in Edinburgh, Birmingham, London, and Paris. We are grateful to the Thomas Fisher Rare Book Library at the University of Toronto for permission to reproduce some of the charts, in our Introduction, from its copies of the *Atlas* and *Breviary*. We are also grateful to the United States Library of Congress's Rare Book & Special Collections Reading Room, for providing the chart reproduced as Figure 4 of our introduction.

Ian Spence is deeply grateful to the late Sir Edward Playfair for his encouragement and many suggestions and criticisms; to Hugh Playfair for his invaluable genealogical research on the Playfair family; to John Lawrence Playfair for lending an original autograph of William Playfair; to Professors Antoine de Falguerolles, Michael Friendly, Gilles Palsky, and Hugh Torrens, and the late Professor John Fauvel for wise criticism and words of encouragement.

We are also grateful for the support, comments, and suggestions of a host of colleagues – too numerous to name – who take an interest in the history of statistical graphs.

Final thanks go to Lauren Cowles and the staff of Cambridge University Press. Their vision and sense of history inspired them to produce this high-quality volume.

Howard Wainer
Ian Spence

Introduction

Sometime in 1787, just two years before France was plunged into revolution and chaos, the Count of Vergennes delivered a package to the royal court of France for the attention of the king. The gift for Louis XVI had come to Vergennes from Lord Lansdowne, an English politician who was on intimate terms with many in the upper echelons of Parisian society. Vergennes was certain that Louis XVI would be very interested in the contents of the package.

The gift was a book written by a young Scottish engineer and entrepreneur who had recently moved to Paris with hopes of making his fortune. His book had been published in London during the previous year and was entitled *The Commercial and Political Atlas* but, unlike more conventional atlases in this era of great exploration, it contained no maps. It did contain charts, but of a new and unfamiliar variety. Louis XVI, an amateur of geography and the owner of many fine atlases, examined his acquisition with great interest. Although the charts were novel, Louis had no difficulty in grasping their purpose. Many years later, their author wrote that

> [the king] at once understood the charts and was highly pleased. He said they spoke all languages and were very clear and easily understood. (Playfair, 1822–3)

A further indication of the king's approval was the royal permit he granted for the establishment of a factory to work metals in Paris. Playfair had intended to use a steam engine to drive a rolling mill, modeled on the machinery and practices in the Birmingham factory

I

of Boulton & Watt, where he had worked from 1777 until 1781. In addition to his endorsement of the venture, Louis XVI donated the land.

The *Atlas* that had captured the king's imagination contained numerous tables and graphs that summarized trade between England and several other countries, as well as a variety of charts that displayed economic data. In total, the volume contained 44 charts. The king of France had grasped fully the significance and utility of the novel representations in the *Atlas*. Importantly, he had understood the universal appeal of the new diagrams. Of course, the use of tables to present economic data was not new, having been common for more than a century after John Graunt (1620–74), who had used them extensively in his *Natural and Political Observations Made upon the Bills of Mortality*, and Sir William Petty (1623–87), who had examined the role of the state in the economy in his *Treatise on Taxes and Contributions*. Coincidentally, both books were published in the same year, 1662. But the pictorial representation of statistical data was revolutionary. The *Atlas* showed, for the first time, how economic data could be represented by charts. The favorable assessment of the ill-fated Louis XVI – who was to perish under the guillotine less than six years later – was both fitting and prescient. A century and a half later, in 1937, the great American historian of statistics, H. G. Funkhouser, echoed the sentiment of the king when he said that "the graphic method is rapidly becoming a universal language."

Today there is scarcely a field of human activity that does not make use of statistical charts like those in the volume delivered to the king of France. The invention can lay fair claim to being one of the most versatile and useful tools for analyzing and displaying data in the sciences and humanities, in commerce and the arts, and in everyday activities that affect us all. Graphs convey comparative information in ways that no tables of numbers or written accounts ever could. Trends, differences, and associations are seen in the blink of an eye. The eye perceives instantly what the brain would take seconds or minutes to infer from a table of numbers, and this is what makes graphs so attractive to scientists, business persons, and many others. The charts allow the numbers to speak to all, and they transcend national boundaries – a Chinese can read the same graph that a Russian draws. There is no other form of human communication that more appropriately deserves the description "universal language."

WILLIAM PLAYFAIR

The author of the *Atlas* was no ivory-towered theoretician. William Playfair was trained as a practical engineer by giants of the Industrial Revolution. Although a craftsman by trade, he was also exposed to the best academic minds of the Scottish Enlightenment, which has so profoundly helped to shape our modern world (Broadie, 2003; Buchan, 2003; Herman, 2001). William was born on 22 September 1759, in the small rural village of Liff, near the city of Dundee. He was born a twin, the fourth child, in the family of the Reverend James Playfair, a Presbyterian minister of the Church of Scotland. Sadly, the twin brother, Charles, like too many other children in those days, did not survive to see his first birthday. In the early years, the Playfair children were educated at home by their father. However, upon the relatively early death of James Playfair, when William was just 12, the role of teacher was thrust upon the eldest brother John, then 24. John would soon gain worldwide fame as a mathematician, physicist, and geologist and would become one of the most distinguished professors at the University of Edinburgh. William Playfair was raised and educated in the presence of genius.

John's scientific approach was unequivocally empirical; one task that he gave his younger brother was to keep a graphical record of daily temperatures. Many years later William acknowledged this childhood exercise as the inspiration for his economic time series line chart. Additionally, and also significant for his intellectual development, John introduced William to many of the great figures of the Scottish Enlightenment, such as the philosopher Dugald Stewart and the economist Adam Smith. John also commended his brother to William and Robert Small, educators who were exceedingly well connected in the 18th-century world of letters, science, medicine, and politics. The Small brothers would play a crucial role in the future training of William Playfair.

The Rev. Robert Small and the Rev. James Playfair were well acquainted, being fellow ministers in nearby parishes in the Presbytery of the city of Dundee. Both had received their Doctor of Divinity degrees from St. Andrews University, the oldest in Scotland, and both were enthusiastic teachers; they had many interests in common. Robert's brother, Dr. William Small, was trained as a natural philosopher and physician at Marischal College, Aberdeen. In 1758 he joined the faculty of William and Mary College in Williamsburg, Virginia, where he served as a professor of mathematics and natural philosophy for six years. By Thomas

Jefferson's own admission, William Small was his most important men-
tor at William and Mary, and the pair maintained a close private friend-
ship and correspondence through the difficult years surrounding the
American Revolution. In his autobiography (1821), Jefferson wrote:

> It was my great good fortune, and what probably fixed the des-
> tinies of my life that Dr. Wm. Small of Scotland was then professor
> of Mathematics, a man profound in most of the useful branches of
> science, with a happy talent of communication, correct and gen-
> tlemanly manners, and an enlarged and liberal mind. He, most
> happily for me, became soon attached to me and made me his
> daily companion when not engaged in the school; and from his
> conversation I got my first views of the expansion of science and of
> the system of things in which we are placed. (4)

On his return to Britain, William Small became a founding member
of the Lunar Society of Birmingham (Schofield, 1963; Uglow, 2002). In
many ways, he was the central figure of the group, and his early death at
41 caused great distress to James Watt and the other members.

At the age of 14, William Playfair left the family home to apprentice
with Andrew Meikle, a well-known Scottish engineer and the inventor of an early threshing machine. Meikle was miller and millwright to
the Rennie family, owners of the Phantassie estate at East Linton, near
Edinburgh. One of the Rennie boys, John, also worked at the Houston
Mill under Andrew Meikle's instruction during the same years that
William Playfair served his apprenticeship. John Rennie would later
become the renowned engineer responsible for the London, Waterloo,
and Southwark Bridges, as well as several other significant engineering
structures. After three years with Meikle, William was recommended by
Robert Small to the position of draftsman and assistant to James Watt,
during the early days of the Birmingham steam engine factory.

James Watt (1736–1819) ranks among the most famous of all engineers. This consummate craftsman and scientist did not build the first
steam engine, as is so often popularly supposed, but there is no doubt
that his improvements converted a primitive, balky, awkward, and inefficient device into the workhorse of the Industrial Revolution. His development of the Newcomen engine was so successful that for all practical
purposes we may say that Watt did "invent" the steam engine. His most
important contribution, in 1765, was the separate condenser, which he
included in his first patent of 1769. The work was largely completed at
the University of Glasgow, but it was not until 1776 that the first practical

engine was built, and the construction of such engines did not become routine until the mid-1780s. Had it been left to Watt, a man subject to despondencies, his momentous ideas might never have come to full fruition and achieved great commercial success. The eventual success was based on Watt's collaboration, starting in 1774, with Matthew Boulton, who had established, in the Midlands city of Birmingham, an engineering factory called the Soho Manufactory that became world famous for its organization and novel equipment. It was William Small who introduced Boulton to Watt and who had encouraged the partnership that led to the development of the steam engine manufacturing company, Boulton & Watt, which was to revolutionize work throughout the world in the most fundamental way.

William Playfair arrived in Birmingham, England, in 1777. He worked in a variety of capacities at Boulton & Watt, but one of his most important duties was as draftsman and clerk to James Watt himself. Watt did not spend much time at the Soho Manufactory, preferring to work alone in his house at Harper's Hill. It was there that Playfair helped Watt with his engineering drawings, although Watt, who was always a demanding critic, did not have the highest opinion of Playfair's drafting skills, referring to him as a "blunderer" in a letter to Boulton in 1778. When it came to patent applications Watt prepared his own drawings, apparently because he was less than pleased with Playfair's efforts. Nonetheless, Playfair continued in this post until the autumn of 1781, so Watt cannot have been completely dissatisfied with his work. Indeed, Boulton indicated, in a letter to Watt, that he was sorry that Playfair was leaving, since Watt would no longer have his assistance in drafting. Blunderer or not, Playfair's experience in drafting and printing drawings for Watt would later serve him well when he turned his hand to writing.

During his time in Birmingham, Playfair became acquainted with several members of the Lunar Society. This distinguished group of businessmen and scientists included Boulton, Erasmus Darwin, Edgeworth, Keir, Priestley, Watt, and Wedgwood. The unusual name of the society derived from the meeting time of the group – they met monthly, from 1765 until 1813, on the Monday evening closest to the full moon so that there would be sufficient light for the late night walk home. The Lunar Society was second only to the Royal Society as an important meeting place for scientists and inventors. Its members were interested in more than pure science – they were passionately engaged in the application of new ideas in natural philosophy to manufacturing, mining, transportation, medicine, and education. The members of the Lunar Society were

the heart and soul of the Industrial Revolution, and they were convinced –
with much justification – that they were changing the world for the better.
Thus William Playfair was privileged to be at the cutting edge of science
and industry in a Britain that was to dominate the world in the century
following the Industrial Revolution – a revolution that was spawned in
late 18th-century Birmingham. Playfair rubbed shoulders with the leading figures of the day in science, engineering, business, and politics, and
they, unknowingly, helped to shape his statistical creations.

In 1779, while still at Boulton & Watt, Playfair married Mary Morris,
and in 1780 their first child, John, was born. A draftsman's wages may not
have seemed adequate to support the new family, and as soon as Playfair
felt he had learned enough to strike out on his own, he left Boulton
& Watt in 1781 with a fellow Soho employee, William Wilson, to form
a silversmithing business in Marylebone, London. From the start, the
new venture was plagued by disputation and bad debts. In a pattern
to be repeated many times in the coming years, Playfair had embarked
upon a speculative grand scheme that was doomed to failure. It seems
that his reach always exceeded his grasp. Despite obtaining four patents
for devices to fashion metal objects, from silver trays to horseshoes, the
business was not successful and Playfair turned his hand to writing.

Playfair's developing interest in writing about economics was in-
tensely practical. As Andrew Meikle's apprentice and James Watt's draftsman, Playfair had been a first-hand witness to the work of several great
engineer-entrepreneurs, including not only Meikle and Watt, but also
Matthew Boulton, John Rennie, Josiah Wedgwood, and James Keir. He
had observed the development and success of Boulton's manufactory
at Soho, the world's first factory to be organized and run in ways that
we would recognize today. Playfair's first publication on economics appeared in 1785, but it contained no charts. A preliminary edition of the
Atlas, with engraved charts, also appeared in 1785 – this was privately
circulated to a select few for criticism. The *Commercial and Political Atlas*
of 1786 was the first publicly available volume to contain charts, and it
exhibits 43 variants of the time series line graph together with a solitary
bar chart. Playfair issued a second edition, which was little changed, in
1787. Despite isolated critical approval, this foray into publishing made
neither riches nor reputation for Playfair and he left England in 1787 to
seek his fortune in Paris. British industry and commerce were leading the
world, and Playfair believed that with his experience at Boulton & Watt
he would be well placed to profit in a France striving to industrialize and
catch up to her neighbor and traditional enemy.

Playfair planned to establish a rolling mill in Paris and, for the moment at least, it seemed that writing was to take a back seat to engineering. Although the plan was approved by Louis XVI himself, it appears that the venture never got off the ground because Playfair was soon involved in other speculative schemes. One ambitious project was the débacle subsequently known as the Scioto speculation. This was a complicated, fascinating, and murky business originating far beyond the borders of France, in postrevolutionary America, and although much has been documented by historians of early American corporations (Belote, 1907), many of the crucial details remain shrouded, including Playfair's precise role in the collapse of the scheme. At the end of the Revolutionary War, syndicates were formed to purchase large blocks of land and to sell individual tracts at an advanced price to European settlers. The American Scioto Land Company established a branch in Paris to peddle the idea to minor French aristocracy, many of whom were becoming increasingly uncomfortable in the rapidly changing political climate of 1788–9. Joel Barlow, the unilingual American representative of the Scioto Land Company in Paris, needed an English-speaking partner who was familiar with the language and local customs. Playfair fit the bill. Although large sums were subscribed and several hundred French citizens emigrated to the wilds of Ohio, the venture ultimately failed and Playfair was accused of hastening the collapse by embezzling funds. However, mismanagement on the part of Barlow and Playfair, coupled with the unpreparedness of the early settlers in a difficult environment, seem equally likely reasons for the failure. Scioto was not the only problematic speculation to occupy Playfair during his years in Paris; he was involved in other legal and financial entanglements and was forced to leave France shortly before the Terror of 1793.

He spent the years between 1793 and 1814 in London, with occasional excursions to the Continent. During this time, he published several books that included charts, the most notable being *Lineal Arithmetic* (1798), the *Statistical Breviary* (1801), and *An Inquiry into the Permanent Causes of the Decline and Fall of Powerful and Wealthy Nations* (1805). In 1809–11 he published the illustrated *British Family Antiquity Illustrative of the Origin and Progress of the Rank, Honours, and Personal Merit, of the Nobility of the United Kingdom*, which included chronological diagrams; hopes of substantial subscriptions from the aristocracy were evidently the motivation behind this mammoth nine-volume endeavor. In business, always seeking new ways of making money, he attempted to import some of the freewheeling financial schemes that he had used in Paris, but the

Bank of England was even less tolerant than the French authorities and Playfair narrowly escaped prosecution in 1797. He continued to write – his output numbering more than a hundred books and pamphlets – but without great monetary success. His many writings on economics include a critical edition (1805) of Adam Smith's *An Inquiry into the Nature and Causes of the Wealth of Nations*. Smith's admirers thought Playfair's additions and commentaries to be insufficiently respectful, and the edition was not well received. In general, his political views were expressed in typically brash and forthright fashion, and his candor did little to win him friends. He coedited a daily paper, the *Tomahawk*, and also a weekly, *Anticipation*, but both soon failed. He frequently fell back on his engineering training, working as a gun-carriage maker, and from time to time he supplemented his income by dubious means. One swindle led to conviction at the Court of King's Bench in 1805.

When the Bourbon monarchy was restored, Playfair returned briefly to France. In 1789, he had initially been in favor of the revolutionaries, but their later excesses forced a change of mind and his subsequent royalist views, which he was never shy to express, made him unwelcome during the years of the republic and empire. In 1814, after the accession of Louis XVIII, Playfair felt that he might, once again, seek better times in Paris. He was appointed editor of an English language newspaper, *Galignani's Messenger*, and he wrote a number of pieces on the state of France. He does not seem to have engaged in business affairs of any consequence, and he eventually fled the country after being convicted of libel. Once more, his reckless and outspoken opinions had constrained his options.

Playfair lacked money in his final years. Writing did not produce the anticipated income and, in worsening health, he lost his enthusiasm for the grand scheme. Although his two sons were independent by this date, life was not easy for a man who was supporting a wife and two daughters, one of whom was blind. In 1816, short of cash, he descended to attempted extortion when he tried to sell some papers alleged to relate to the great Douglas Cause of half a century earlier. Lasting seven years, the Cause had been the longest and most expensive legal proceeding in Scottish history. The documents that Playfair offered to Lord Douglas were relevant to the alleged imposture of newborn twins conducted in Paris many years previously. The papers were said by Playfair to have cast doubt, yet again, on the legitimacy of the Douglas inheritance. These papers almost certainly never existed; they were merely a prop in Playfair's plan to extort money from one of the richest men in Scotland. Because of Douglas's resistance and Playfair's weak evidence, the blackmail did not succeed and indeed did not come to light until recently (Spence & Wainer, 1997). This shameful affair demonstrates Playfair's straitened financial situation and his readiness to ignore the law when it suited his purposes.

The last few years of Playfair's life saw a renewed interest in economics, and Playfair's final publications include some very fine charts (Playfair's two letters on agricultural distresses, 1821, 1822). These late works examined the difficulties experienced by English farmers in the early 19th century. Playfair died on 23 February 1823 in Covent Garden, likely in the house at No. 43 Bedford Street. He was survived by his wife and four of his children, one of whom, Andrew William, had emigrated to Canada where he was prominent in the military and successful in private business, eventually founding the town of Playfairville not far from the capital, Ottawa. Andrew William persuaded his older brother John to join him, and their descendants have prospered and spread throughout Canada.

During his life, despite the interest and approval of a select few, William Playfair's invention of statistical graphs went largely unacknowledged. Although he was a tireless advocate for his charts, he made few converts. His obituaries ignored the graphical inventions and concentrated on his political and economic writings, which were not held in high regard by his contemporaries, although they have attracted renewed interest today. One apologist wrote:

> Had Mr. Playfair cultivated his mechanical genius, there is no doubt, that he would not only have obtained considerable eminence, but have rendered no inconsiderable service to this country. Unhappily, however, for his own interests, he had the ambition to become an author. (Author unknown, *Edinburgh Annual Register*, 1823, 332)

18TH-CENTURY BARRIERS TO STATISTICAL CHARTS

Today, most people think that statistical graphs are such simple and obvious creations that almost anyone could have invented and published them. Indeed, it is their simplicity that accounts for much of their appeal, and that is why we give scarce thought to the ingenuity required to invent and promulgate statistical charts. Familiarity has dulled our appreciation of their significance, diminishing the importance of their creator, whose name until recently was largely unknown, even to professional statisticians. But the idea of devising and publishing statistical graphs was not obvious two centuries ago and, even today, the form is not nearly so naive and self-evident as it might first appear (Cleveland, 1985; Tufte, 1983; Kosslyn, 1994; Spence & Lewandowsky, 1990; Wainer & Velleman, 2001; Wainer, 2000, 2005).

Large collections of economic statistics were widely available – and had been since the time of Graunt and Petty – more than a century before Playfair thought of publishing such data in pictorial form. The data necessary for the invention of statistical graphs were present in abundance, but no one else had the inspiration to represent them as pictures. There were various impediments to the publication of illustrations in serious writing. There were philosophical objections, concerns regarding accuracy and misrepresentation, and technical barriers to publication.

Plants have been portrayed in print since the introduction of the printing press in the 15th century. From early Renaissance herbals, through pictures of Baroque gardens, to increasingly naturalistic depictions of plants and flowers in the 17th and 18th centuries, printed illustrations of natural history had become fairly common and accepted. But there is ample evidence to suggest that similar illustration in serious scientific writing was viewed with suspicion, and eminent experimenters like Robert Hooke, who used illustration, did so with misgivings. About his *Micrographia* of 1665, which contained many illustrations, Hooke wrote that

> Pictures of things which only serve for ornament or Pleasure, or the Explication of such things as can better be describ'd by words is rather noxious than useful, and serves to divert and disturb the Mind, and sways it with a kind of Partiality or Respect. (64)

Hooke worried about the possibility of misrepresentation and took great pains to assure the reader of accuracy, or to point out possible distortions that the illustrations might produce in the reader's mind. Tilling (1975) has pointed out that in the 17th century information in charts produced by automatic graphical recording devices, such as weather clocks, was often translated into tabular form and that, with one exception in 1724, there was no publication of similar charts until the 19th century. Presumably no value was seen in graphical presentation or, more simply, the continuous graphical record was not regarded as being as trustworthy or informative as the corresponding sequence of numbers.

Biderman (1990) and Valois (2000) have argued that a mistrust of sense perception on the part of Descartes and his disciples was a powerful impediment to the development of empirical methods of investigation, including the development and use of statistical graphs. This impediment was to disappear as British empiricist philosophy evolved during the late 18th century. Beginning with Locke, Berkeley, and Hume, British empiricists in the 17th and 18th centuries had argued that knowledge comes from experience whereas the rationalists, such as Descartes, had maintained that knowledge may be derived solely through reason based on innate ideas. The empiricists rejected the notion of innate ideas and argued that almost all knowledge is based on sensory experience. The Scottish realist philosophers, such as Thomas Reid and Dugald Stewart, took this idea even further by emphasizing the process of inductive reasoning from sense data. In a distinction that still has currency today, Reid (1764) characterized the difference between rationalist philosophy and empiricism in two ways. Firstly, rationalists believed that concepts are known intuitively through reason, as opposed to experience. And secondly, rationalists maintained that truths could be deduced from innate ideas, in much the same way that theorems are deduced from axioms. Mathematical proof was the model for obtaining knowledge. Although the Scottish empiricists also used deductive reasoning when appropriate, they attached much greater importance to the inductive method. Both Thomas Reid and Dugald Stewart, Reid's intellectual successor, rejected much of Cartesian philosophy, preferring to rely on observation and inductive reasoning. Although more than two centuries old, their approach is surprisingly modern. Nowadays, we take for granted that empiricism is at the core of modern scientific method, which considers that theory should be tested by observation rather than by intuition.

The Playfair brothers were well acquainted with the leading empiricist philosophers of the Scottish Enlightenment, such as Hume, Reid, and Stewart, and their own thinking was in the same empirical mold. John Playfair and Dugald Stewart were firm friends and colleagues and their approaches to mathematics and natural philosophy were highly compatible. Indeed, in 1785, John Playfair succeeded Stewart as professor of mathematics when the latter relinquished his chair to take Adam Ferguson's chair in moral philosophy. As a boy, William Playfair had absorbed the prevailing Scottish empirical approach in science, learning to represent physical data by line graphs under the instruction of his brother John (Playfair, 1805, xvi). However, although such charting had become common among natural philosophers for tracking their private experimental data, they did not employ these devices to buttress arguments via publication. John Playfair, for example, never made use of the line graph in his writings (J. Playfair, 1822). It was William Playfair's genius not only to apply the line graph to economic data but to see the value in publication. However, despite the shift to empiricism, a general mistrust of pictorial representation persisted, with 18th- and early 19th-century academics reluctant to publish graphs of physical or statistical data, possibly because of lingering concerns regarding accuracy, or simply the sheer technical difficulty of publication.

COPPERPLATE ENGRAVING

A significant barrier was the process of printing illustrations. As 'Espinasse (1962) has noted, in the 17th century, scientists were versatile, used to manufacturing their own apparatus and to dealing with tradesmen and craftsmen, whereas by the 18th century, scientists had become more specialized and were less likely to possess the practical expertise to engrave their own plates. But William Playfair did have the skills. He was no academic insulated from the real world; his training had prepared him perfectly for the invention and production of statistical charts. Mastering the printing process was easy for Playfair, the engineer and draftsman, and we know that he even frequently engraved the lines on the copper plates himself, leaving the more delicate work of lettering and decoration to the printer.

In the 18th century, intaglio copperplate prints of maps or natural history were usually signed by the artist and the engraver. The artist's name often appeared in the white space beneath the image on the left-hand side, followed by "del.," an abbreviation for the Latin *delineavit*, and the engraver/etcher's name would appear on the right side, with "sc.," an abbreviation for *sculpsit*. In most of Playfair's plates one engraver's name, Neele, appears in the traditional position on the lower right, but the early plates by J. Ainslie are signed in the lower left. Playfair's first engraver, John Ainslie (1745–1828) was also an outstanding cartographer, surveyor, and publisher. Born in Jedburgh in the Scottish Borders, he is best known for "Ainslie's Travelling Map" (1783) and his large-scale map of Scotland (1789). Nine of the first 10 plates in the first edition of the *Atlas* are by Ainslie. One of the plates is unattributed and, judging by the small differences in style from Ainslie's plates, it may have been completed by Playfair himself. The first 10 plates are dated 1785 and may have been the only plates to figure in the preliminary edition that was privately circulated to Watt, Boulton, and a few others. All except one of the remaining plates, dated 1786, are by Samuel John Neele (1758–1824), a London engraver and copperplate printer who specialized in book illustrations and maps. Neele signed most of his engravings using only his surname and the address of his business at 352, Strand, London. Both engravers were craftsmen of great ability, and many fine examples of their maps still survive. However, the quality of engraving in Playfair's charts is not comparable to the excellence of the cartography produced by these two men in their customary work. The likely explanation is that Playfair required his work to be done economically and requested the craftsmen to minimize expensive frills.

A copperplate engraving is an image taken from an engraved copper plate. Although copperplate printing developed as early as the 14th century, it was only during the 17th and 18th centuries that copperplate engravings were widely used for illustrated works in France and England. Copperplate remained the standard until the early 19th century when engraving on more durable steel plates took over. A plate of bright, burnished copper is first coated with a ground, usually a hard wax, and then the desired image is traced with a needle. The ground is then removed. Guided by the traced lines, the engraver uses a burin, a metal tool with a sharp point, to engrave onto the copper plate. Metal shavings are cut away by the burin. These shavings, or "burrs," must be detached by a "scraper," another cutting tool. The deeper the cut, the stronger the printed lines will be. The plate is then warmed, inked, and passed through a press with the sheet of paper to be printed.

Etching was also used to produce the design, which was first drawn with a needle to penetrate an acid-resistant wax coating previously applied to the plate. The plate was then immersed in a bath of acid, which etched those areas exposed by the needle. Deeper lines were produced by re-immersion in the acid, after lines that were not to be deepened had been stopped out with acid-resistant varnish. Without a microscope it is difficult to distinguish engraving from etching, with the latter producing slightly fuzzier lines. Most illustrators in the 18th century employed a combination of the two processes. Neele and Ainslie probably used both methods to create the decoration, framing, titles, and other lettering, but in many cases Playfair probably engraved the data lines himself. Several plates show evidence of careless or inexpert use of the burin to engrave the data lines and the grid lines; in the third edition, reproduced here, there are obvious blunders in a few plates and minor errors in several others.

Because copper is a soft metal, it was rare to take more than 1000 impressions before the plate deteriorated. This is one of the reasons why many works at the time appeared in several editions soon after the original. These new editions provided an opportunity to add new information, as Playfair did with the third edition of the *Atlas*.

THE COMMERCIAL AND POLITICAL ATLAS (1786–1801)

In 1785, Playfair first wrote about economics in *The Increase of Manufactures, Commerce, and Finance, with the Extension of Civil Liberty, Proposed in Regulations for the Interest of Money*; this extended essay contained no charts. In the same year, he also circulated a version of the *Commercial and Political Atlas* to selected acquaintances. He sought criticism that would help him to improve the published version. He sent copies to both Watt and Boulton, writing to the latter:

> I have taken the liberty of asking if you will do me the favour to accept of the first number of a work which I hope may be one day of considerable utility. I must beg leave as a particular favour that you will oblige me with your remarks and any hints for amendment or improvement that may occur to you as I wish to make it as perfect & complete a work as I am able. (20 September 1785)

A similar letter must have been sent to Watt who replied:

> I can think of nothing in addition to your plan, except that it might be proper to give in letter press the Tables from which the Charts have been constructed . . . for the charts now seem to rest on your own authority, and it will naturally be enquired from whence you have derived your intelligence. A general chart showing the increase & decline of various articles of exportation & importation, or particular charts of the same would also be useful, but my mind is so much turned to other pursuits that I cannot pretend to direct you. (10 October 1785)

Watt, ever the careful scientist and engineer, was concerned with the accuracy and the provenance of the data and Playfair, who revered Watt, was persuaded to include tables in the first (1786) and second (1787) editions of the *Atlas*. By 1801, Playfair had decided to ignore Watt's advice, and no tables are to be found in the third edition.

The bulk of the *Atlas* examined English commerce with other nations during the 18th century. The first edition was published in foolscap folio, cut in the size of 8 × 13 inches (216 × 330 mm), with the type set in landscape format rather than in the more common portrait format. The charts appear on separate pages at the same size as the text. From about the 16th to the 18th centuries, illustrations were printed using engraved copper plates, inked and wiped so that the ink remained only in the incisions. However, this *intaglio* method was at odds with the printing of text where the type was in relief. Consequently, the cheapest solution was to print the illustrations separately (Biderman, 1981; Spence, 2000). Playfair was probably driven to the less common landscape format to accommodate charts that would have been much less successful if they had been constrained to a vertical format. In using the more effective landscape layout, Playfair seems to have anticipated modern ideas regarding the optimal aspect ratios of graphs (Cleveland, 1985). Although the second edition uses an identical layout, by the third edition, Playfair had adopted a more conventional vertical format for the text. Partly because of this change, some charts, which are two to three times the page size, appear as folded flyouts from the main body of the work. Playfair could have achieved the same end by printing these charts sideways – and he did this in most instances – but he clearly wanted to be able to present the more important charts with sufficient detail.

In the first edition, 43 of the 44 charts plotted pounds sterling on the ordinate against time on the abscissa. In addition, a solitary bar chart – an oddity made necessary because Playfair did not have sufficient data to construct a line graph – was the only chart that did not include time as a dimension. The bar chart was directly inspired by Priestley's (1765) chronological charts and Playfair must have been aware of the irony when he apologized for the anomaly: "This Chart . . . does not comprehend any portion of time, and it is much inferior in utility to those that do" (*Atlas*, 1786, 101).

The graphs in the three editions of the *Atlas* were remarkably similar to those in use today; hachure, shading, color coding, and grids with major and minor divisions were all introduced in the various editions of the *Atlas*. Actual, missing, and hypothetical data were portrayed, and the kind of line used, solid or broken, differentiated the various forms. Playfair filled the areas between curves in most of the charts to indicate accumulated or total amounts. All included a descriptive title either outside the frame (as in the first edition) or in an oval in the body of the chart (as in the third edition). The axes were labeled and numbered where the major gridlines intersected the frame.

DIFFERENCES BETWEEN THE EDITIONS

There are few differences between the first and second editions. The first edition contains 44 charts in 40 plates – copperplates colored by hand in three or four colors. Twenty of these charts represent trade between England and other countries. The line of imports is stained orange-yellow and that of exports is stained red. The area between the curves is a bluish-green when the balance is favorable to England and pink otherwise. In the second edition, the charts differ mainly in the loss of color.

Before the invention of multiple-layer color printing, printed black ink intaglio illustrations were sometimes colored freehand. Water color was applied to the image to enhance the design in much the same fashion as in a watercolor painting. By the 18th century copper engravings colored by hand were fairly common. Colorists were employed by the artist or publisher to produce the finished color prints although, in Playfair's case, there is reason to believe that he colored many of the charts himself to save money. To color or not to color forced a choice on the engraver. An engraving that was not to be colored had to represent color and shape by means of the engraving alone. Hues were represented by cross hatching or stippling with dots. Outlines were generally engraved with a heavier hand, resulting in a print with the look of a pen and ink drawing. On the other hand, copperplate engravings that were to be hand-colored depended on the colorist to add detail and shape. Engraved lines were kept light so as to not interfere with the water colors. The heyday of hand-coloring lasted for about a hundred years, from the middle of the 18th to the middle of the 19th centuries, when new color printing processes became available.

Playfair moved to France in 1787 and probably had neither the time nor the money to do the laborious hand-staining that was necessary to color the charts. He replaced the two major colors, representing favorable and unfavorable balances, by using hachure (for blue-green) and stippling (for pink). Otherwise, essentially the same charts appear in both editions and also in *Lineal Arithmetic* (1798), where color, hachure, and stippling are all used, perhaps an indication of slightly easier fiscal circumstances for Playfair.

There are much more substantial differences in the third edition. Instead of 40 plates containing 44 charts, there are 28 plates containing 33 charts. The most significant omission is the bar chart showing the exports and imports of Scotland. Gone also are charts showing trade data for England, Holland, and the United States. Three charts showing aspects of the national debt as it related to annuities and interest rates were also dropped. Missing also are the five charts, attributed to James Corry, representing economic data from Ireland. Two new charts were added. The first, in Plate 19, is a rather elaborate large area chart on a flyout showing the annual revenues of England and France as well as the interest on debt. This chart includes a chronological display at the top which shows the reigns of English, British, and French monarchs. The other new plate is not numbered, although it is given a figure of 26 in the index and referred to as Chart XXVI in the text.

Although much of the data are brought up to date, the tables of numbers are no longer incorporated. James Watt had advised their inclusion in the earlier editions to allay possible concerns regarding provenance or accuracy. In fact, as we shall see, the tables call attention to Playfair's lack of concern for accuracy, and so Watt's wise counsel had an effect opposite to what he intended. Playfair's goals were didactic and at times polemical, rather than analytical, and his freehand drawing of the variations in imports or exports is sometimes hard to reconcile with the numbers. While he certainly made small errors and technical mistakes, the most egregious problems concern his interpolations between data points. On many occasions, the ups and downs of the lines are fanciful and probably reflect Playfair's prejudices rather than the likely values of the missing data.

Playfair did not believe that the accuracy of a graph could exceed that in a table. Nor did he feel that such accuracy was necessary:

> The advantage proposed by this method, is not that of giving a more accurate statement than by figures, but it is to give a more simple and permanent idea of the gradual progress and comparative amounts, at different periods, by presenting to the eye a figure, the proportions of which correspond with the amount of the sums intended to be expressed. (*Atlas*, 1801, ix–x)

Nevertheless, it is surprising that he was not more concerned with accuracy since he was well aware that his new methods might arouse suspicion. He tackles the issue head-on in the *advertisement* that opens the first edition:

> As to the propriety and justness of representing sums of money, and time, by parts of space, tho' very readily agreed to by most men, yet a few seem to apprehend that there may possibly be some deception in it, of which they are not aware ... (*Atlas*, 1786, iii)

As we have noted, James Watt also had stressed the importance of accuracy, and he persuaded Playfair to include at least some of the original data in tabular form. But, alas, despite Watt's counsel, the *Atlas* is not a model of precision. Several arithmetical errors and careless drawing are evidence of rushed production. Some of the curves that connect the data points seem to have derived their shapes from Playfair's opinion of how the intervening data should look. His curves are drawn freehand, often somewhat crudely, betraying a lack of practice in the demanding skills of engraving. We give some examples of these peculiarities and technical flaws:

- *Arithmetic.* In the section labeled "Contents of the plates in numbers" (first ed., 1786, 21–2) there are minor, but careless, arithmetical errors in the tables. Several of the "balances" fail to agree with the difference between "exports" and "imports." These faults, in themselves, are trivial, but such carelessness throws into doubt the accuracy of the remaining numbers in these and the other tables.
- *Plate 4.* This plate, which has the same number in all editions of the *Atlas*, illustrates several problems. While Playfair's rendering of the raw data (Table IV in the first edition) is reasonably faithful in the first and second editions, there are large discrepancies between the numbers and the curves in the third edition. In our Figure 1, we have superimposed smoothed blue (import) and green (export) curves that are consistent with the tabulated data. There are substantial differences between our curves and Playfair's depiction, which is also considerably different from the graph that he had published 15 years before. A possible explanation is that Playfair had discovered more accurate data in the interim – if so, he does not allude to this in the text. Since there is little doubt that scholars in the 18th century were concerned about the misrepresentations that pictorial methods might bring with them, Playfair may have damaged his case for adoption by constructing charts that did not faithfully reflect the numbers.

  Another curiosity in Plate 4 is that the engraved line for imports terminates around 1785 and, similarly, the line of exports stops at about 1792, although the hand-staining of both lines continues until 1800. There is no obvious reason for these discrepancies, which do not occur in other charts where, if the engraved line terminated early, the color stain usually did not continue beyond that point.

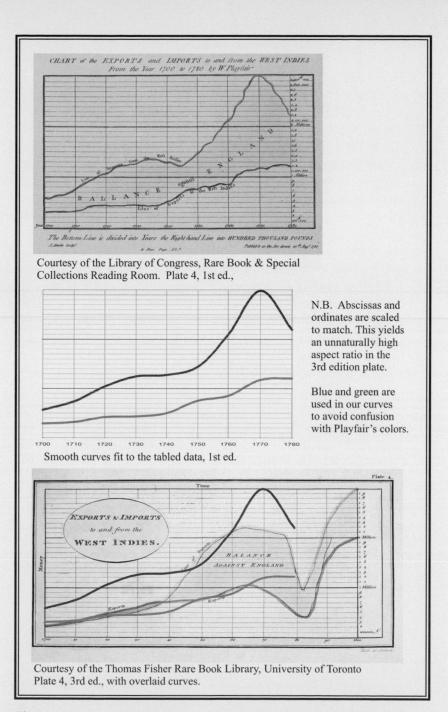

Courtesy of the Library of Congress, Rare Book & Special
Collections Reading Room. Plate 4, 1st ed.,

N.B. Abscissas and
ordinates are scaled
to match. This yields
an unnaturally high
aspect ratio in the
3rd edition plate.

Blue and green are
used in our curves
to avoid confusion
with Playfair's colors.

Smooth curves fit to the tabled data, 1st ed.

Courtesy of the Thomas Fisher Rare Book Library, University of Toronto
Plate 4, 3rd ed., with overlaid curves.

Figure 1 Plate 4 (1st, 2nd, and 3rd eds.) showing trade with the West Indies. We have
matched the scales of the plates to allow a direct comparison of the curves. This has the
consequence of producing a slightly distorted third edition plate that is wider than the
original. The central panel was constructed using the data from Table IV in the first edition
and is discrepant from the curves in the third edition. Note that our reproduction of the first
edition plate is taken from microfilm and lacks the color of the original.

- *Plate 17 (Plate 22 in the first edition)*. This plate, although engraved as number 17, is indexed as Plate 16 in the third edition. The University of Pennsylvania's plate number has been altered by an unknown hand to conform to the index, whereas, as our Figure 2 shows, the University of Toronto plate number remains unaltered. Minor mixups such as this show that Playfair, his engravers, and printers worked at speed, with small errors creeping in. Even today, errors of this sort are not unknown in publishing. However, a more interesting aspect of this chart is Playfair's rendering of the tabulated data into curves. Although his representation does not reflect the actual numbers with scrupulous accuracy, this is a minor problem. It is his conception of how the data have varied during each successive ten-year interval that is curious. As far as we know, Playfair did not attempt to fit the data for each individual year in the several decades – even if these data had been available – and therefore his curve is likely an interpolation. Our conservative smooth interpolation yields just three points where the line of imports crosses that of exports, whereas Playfair's graph exhibits seven such intersections. Thus one gains the impression that the balance of trade oscillated between England and the Channel Islands during each decade of the 18th century, when, in fact, there was probably only one period, from about 1710 until about 1740, during which the balance of trade favored Jersey, Guernsey, and Alderney. There are other plates that exhibit variations in the curves that are not easy to reconcile with the data given in the tables from the first two editions.
- *Plate 5*. Different copies of a work from the same printing show differences. These variations highlight the handmade nature of books two centuries ago. An example is shown in our Figure 3, which reproduces Plate 5 from the third edition of the *Atlas*. The upper plate comes from the copy in the Annenberg Rare Book and Manuscript Library at the University of Pennsylvania and the lower plate comes from the volume in the Thomas Fisher Rare Book Library at the University of Toronto. There are noticeable differences in the handstaining of the two charts. After two centuries, it is difficult to know whether differences in color are a result of differential aging or whether such variation was present in the originals. It seems that the upper version lacked original staining in the frame and title oval. The upper plate has been more hastily colored and the areas between the lines of import and export are much less carefully covered than in the lower plate.

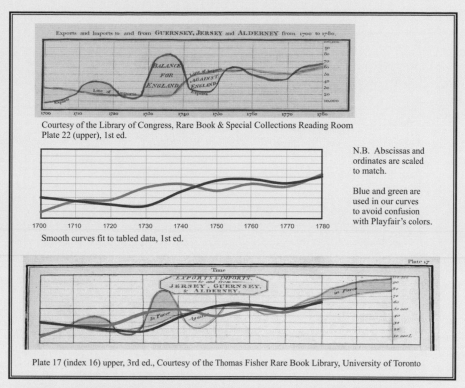

Courtesy of the Library of Congress, Rare Book & Special Collections Reading Room
Plate 22 (upper), 1st ed.

Smooth curves fit to tabled data, 1st ed.

N.B. Abscissas and
ordinates are scaled
to match.

Blue and green are
used in our curves
to avoid confusion
with Playfair's colors.

Plate 17 (index 16) upper, 3rd ed., Courtesy of the Thomas Fisher Rare Book Library, University of Toronto

Figure 2 Plate 17 (3rd ed.) and Plate 22 (1st and 2nd eds.) showing trade
with Jersey, Guernsey, and Alderney. This plate is numbered 16 in the index
to the third edition. We have matched the scales of the plates to allow a
direct comparison of the curves. The central panel was constructed using
the data from Table IV in the first edition. Playfair's curves exhibit seven
crossings, whereas our curves cross only three times. Note that our
reproduction of the first edition plate is taken from microfilm and lacks
the color of the original.

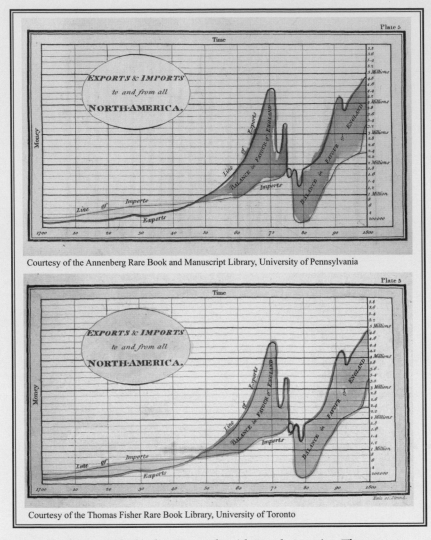

Courtesy of the Annenberg Rare Book and Manuscript Library, University of Pennsylvania

Courtesy of the Thomas Fisher Rare Book Library, University of Toronto

Figure 3 Plate 5 (3rd ed.) showing trade with North America. These two versions of the same plate are reproduced from different copies of the book. Differences in quality and quantity of the hand-staining are noticeable. A common engraving error in the line of imports is seen at the extreme right, with the hand-stained yellow-orange line indicating the correct branch.

There is an engraving error, just before 1800, where the engraved line of imports takes two paths. Playfair probably inscribed the lower branch in error and subsequently added the correct path. The fault is scarcely noticeable since the correct curve is clearly emphasized by the yellow-orange staining.

The engraver's name, Neele, is at the lower right in the University of Toronto's copy but not in the University of Pennsylvania's copy, and there are eight other similar instances. There are also discrepancies in plate numbering for three of the plates and also between the plates and the index. These anomalies suggest that the plate numbering and engraver attributions may not have been engraved on the original copper plate but were added later.

GRAPHICAL INNOVATION

Despite the minor numerical errors, the technical slips, and graphical functions that are occasionally more fanciful than accurate, all three editions of the *Atlas* introduced an astonishing number of novel charting constructions that are still in common use today (Tufte, 1983; Biderman, 1990; Costigan-Eaves & Macdonald-Ross, 1990). We shall comment on only a small sample of Playfair's inventions (some of which are seen in the first and second editions only), and we invite the reader to discover others:

- *The time series line graph.* The third edition has one simple time series line graph (Plate 26). Although all other charts use a line to display variation in amounts, most also include staining or shading between the curves and may be described as area charts. Playfair introduced several different variations of this form.

- *The divided surface area chart.* A good example is found in the first edition (Plate 38, after p. 153). This chart does not appear in the third edition, and is one of several charts that are due to James Corry, after the fashion of Playfair. In the third edition, Plate 19 is a divided area chart. Area charts are ideal for showing trends when the variation in two or more time series must be shown simultaneously. The area between each line and the abscissa is filled with a color for that data series, with the colors in the lower areas occluding those higher up. The use of color (or hachure or stippling) serves to emphasize the ways in which accumulated amounts have varied.

- *The bar chart.* This chart, inspired by chronological diagrams (Funkhouser, 1937; Wainer & Spence, 2005), was introduced in the first edition (see our Figure 4), but by the third edition this chart had disappeared and Playfair did not use the form again.

- *Titles and textual descriptions.* The first edition used descriptive titles above the chart, outside the frame. Explanatory notes regarding the scaling of axes appeared below the frame. Curves and stained areas in the charts were labeled. Other information often also appeared below the frame, for example, the engraver's name, or the date. By the third edition, Playfair had arrived at a more designed – and more expensive – look. The title was relocated to an oval or other shape within the frame; decoration of the text in this caption was common, as in Plates 19 and 21. Our Figures 1 and 2 show this evolution.

- *Framing.* The charts were invariably framed. In the first edition, this consisted of a simple double-lined box, similar to the ones that we have used in our illustrations in this Introduction. By the third edition, the frame included a stained border just inside the double-lined box. This provided a space for the labels and scale values, and it made the chart more pleasing to the eye. Our Figures 1 and 2 provide an illustration of this evolution.

- *Color coding.* Plate 17 provides a good example. A thick red hand-painted line is used for Exports; a yellow-orange line for imports; solid blue-green fill color between the export and import lines shows when the balance of trade is positive – exports exceed imports – and a pink-red solid fill shows when the balance of trade is negative. Thus color is used by Playfair to emphasize the qualitative differences between the time series and the quality and quantity of the varying accumulated amounts.

- *Hachure and stippled dots.* Where color was not available, as in the second edition, Playfair adopted the engraver's practice of simulating dark colors by hachure and lighter colors by stippling.

- *Labeling of axes.* West border: vertical label "Money"; north border: horizontal label "Years"; south border: time scale; east border: money in pounds sterling. This scheme forces the graph to have a frame, unlike common practice today where the labels and scales tend to be on the west and south sides only.

- *Gridlines.* Major gridlines are engraved more heavily than minor gridlines. Plate 1 is a good example. Minor vertical gridlines are not used where there are no data (as indicated in the tables of the first edition). Presumably, after 1780, Playfair had data for each individual year.

Exports and Imports of SCOTLAND to and from different parts for one Year from Christmas 1780 to Christmas 1781

Names of Places

Jersey &c.
Iceland
Poland
Isle of Man
Greenland
Prussia
Portugal
Holland
Sweden
Guernsey
Germany
Denmark and Norway
Flanders
West Indies
America
Russia
Ireland

L 300 000

10 20 30 40 50 60 70 80 90 100 110 130 150 170 200 220 240 260 280

The Upright divisions are Ten Thousand Pounds each. The Black Lines are Exports the Ribbed Lines Imports.

Published as the Act directs June 7th 1786 by Wm Playfair

Neele sculp t 352 Strand London

Figure 4 Plate 25 (1st ed.) showing the trade of Scotland during 1781.

- *Suppression of nonsignificant digits.* Plate 3 is a good example. The scale is implicitly defined by the first label, where full precision is used, thus indicating the value of the intervals between gridlines (however, Playfair is not consistent in his use of this device).
- *Time period indicators.* In Plate 19 of the third edition, the reigns of the kings of England, Britain, and France are shown at the top in the style of the chronological diagram that Priestley used in 1765 to show the life spans of significant personages from classical antiquity. Another example is found in Plate 31 of the first edition (after p.133); the upper three black horizontal bars indicate times of war. In the corresponding Plate 25 in the third edition, Playfair has dropped these bars.
- *Event markers.* In Plates 6, 20, and 21, Playfair uses vertically oriented text in the body of the figure, positioned to mark important historical events that may have had some influence on the subsequent trend in the time series.
- *Theoretical/hypothetical/projected values.* Plate 21 shows the projected reduction in the national debt after a "Sinking Fund" was established by the Prime Minister, Pitt the Younger. The national debt had assumed staggering proportions due to the rebellion of the American colonies, and William Pitt planned to retire the debt by imposing new taxes. But taxation alone was insufficient. In 1786, Pitt, who became Prime Minister in 1783, introduced a Sinking Fund, using an idea first implemented by Walpole in the 1720s. Each year, £1,000,000 of the surplus revenue raised by new taxes was to be added to the fund. The accumulated interest was to be used to pay off the national debt. The system was extended in 1792 so as to take into account any new loans taken by the government. The system worked in peacetime – with regular annual surpluses – but, after the country went to war in 1793, the debt was redeemed by new borrowing at higher rates of interest. The chart in Plate 21 shows Playfair's model for the time course of the reduction in the national debt. Playfair does not explain the mathematical basis for the construction of his curves, other than to allude to compound interest, but he is skeptical of the claims of others, whom he refers to as "arithmetical pedants."

  Although the national debt was a growing concern in 1786, and Playfair discusses both the debt and the Sinking Fund (1st ed., 131–3), his earlier graphical approach was rather different in Plate 30 in the first edition (before p.131). Plate 21 (3rd ed.) and its often-reproduced empirical partner, Plate 20 (3rd ed.), do not appear in the first or second edition.

* *Solid and broken lines.* While most of the lines and curves that Playfair drew are solid, he occasionally used broken lines. Sometimes he seems to have intended this to indicate uncertainty in the data. For example, in Plate 33 of the first edition, he uses a broken line as a companion to a solid line representing the national debt to indicate that "the state of the debt is exact, or very nearly so." Thus the broken line indicates a lower degree of confidence in the accuracy of the data. In the *Statistical Breviary*, broken lines are used to indicate the relationship between population size and taxation (Charts 1 and 2). In this instance there is no suggestion that the data are inaccurate. A broken line is also used with the outer of two concentric circles that represent the area of France and the other territories that it controlled (Chart 2). It is difficult to know whether Playfair used the broken line to indicate a lack of confidence in the data in this case.

THE STATISTICAL BREVIARY (1801)

In the *Statistical Breviary*, Playfair presented statistical data for European countries at the beginning of the 19th century. He used charts since he believed that "making an appeal to the eye when proportion and magnitude are concerned, is the best and readiest method of conveying a distinct idea" (4). The most important graphical innovation in this volume was the pie chart. The intellectual origins of the pie chart remain obscure – although Playfair acknowledged and wrote about the inspirations for the time series line chart and the bar chart, he was silent regarding the pie chart. This diagram seems almost certain to have derived its inspiration from the logic diagrams of Leibniz and Euler (Spence, 2005), but it is likely that pies, circles, and intersecting circles were such simple and familiar forms to Playfair that he did not think explanation or comment was necessary. Nevertheless, his use of pie and circle diagrams, which had until then only been used to illustrate concepts in mathematical logic, was revolutionary. Playfair was a capable and inventive adapter of ideas from other domains (Spence & Wainer, 2001), and his adaptation of logic diagrams to portray and compare empirical data was ingenious.

The first chart in the *Statistical Breviary* depicts European countries before the French Revolution of 1789, and the second chart shows how circumstances had changed by 1801. Circles represent the land areas; for example, Russia, the largest country, is symbolized by the circle of greatest diameter, while small nations like Portugal call for tiny circles. Just below the horizontal diameter of most circles, Playfair has inscribed the values of the areas in square miles. The charts also depict the sizes of the populations and the revenues of the countries, and whether individual countries were maritime powers (area stained green) or nonmaritime powers (stained red). The sizes of the populations and the tax revenues are represented by vertical red lines on the left of each circle and by the vertical yellow lines on the right. The dotted lines joining the tops of the lines show the tax burden on the populations. However, as Funkhouser (1937) noted, "the slope of the line is obviously dependent on the diameter of the circle" and so it cannot serve as an accurate index of the tax burden. Playfair probably intended the reader simply to note whether the slope was positive or negative.

To show how some countries were subdivided, Playfair used several strategies. For example, the Russian empire was divided into European and Asiatic dominions with the former represented by the inner circle and the latter by the surrounding annulus. This diagram uses the two distinct areas to represent the distribution of the empire between the two continents. The Asiatic dominions were represented by the annulus, which was stained green indicating a sea power. The inner circle was stained red to indicate that the European dominions were land powers. The Turkish empire was harder to accommodate since it was spread across three continents: Asia, Europe, and Africa. Three concentric circles would have made visual comparison of the areas even more difficult than in the case of the Russian empire. Playfair's whole purpose in creating these diagrams that represented quantity by area was to make comparisons effortless and the data more memorable. Playfair, who appears to have recognized and understood the perceptual issues involved, realized that concentric circles yield areas whose sizes are hard to compare accurately. Accordingly, he divided the circle representing the Turkish empire into three sectors proportional to the Asiatic, European, and African land areas. Again, he used colored stains: green to signify maritime power (the Asian sector); red to denote land power (the European sector); and yellow (the remaining African sector). Playfair gave no rationale for his use of these particular colors, but this diagram was the first pie chart to display empirical proportions and to distinguish the component fractions by color.

Playfair introduced three new statistical diagrams in the *Breviary*: the circle chart, the pie chart, and a figure to show joint properties, similar to a Venn diagram. Like the line and bar charts, introduced 15 years earlier, his designs have still not been materially improved upon in 2005.

PLAYFAIR AND THE PSYCHOLOGY OF GRAPHS

During the past two decades, cognitive science has played an important role in advancing our understanding of the power and utility of statistical graphs. Graphs achieve their success by capitalizing on the basic perceptual and cognitive capacities of human beings. However, interest in the psychological aspects of charts is not new. Indeed, Playfair seems to have well understood that our cognitive and perceptual capacities were critically important (Costigan-Eaves & Macdonald-Ross, 1990). Perhaps we should not find this surprising since the Playfair brothers were well acquainted with the ideas and methods of the Scottish empiricist philosophers – in particular Hume, Reid, and Stewart – whose enquiries focused on questions in perception and cognition that continue to occupy experimental psychologists.

Playfair believed that graphs would be a powerful aid to memory; intuitively, he appreciated that visual memory was more robust than memory for words or numbers. When he was searching for a better way of presenting tabular data he said (*Atlas*, 1801, xiv), "a man who has carefully investigated a printed table, finds, when done, that he has only a very faint and partial idea of what he has read." He also appreciated that intuitive visual comparisons of size could be made much more rapidly, and almost as accurately, than by mental arithmetic computations, using the numbers themselves. He claimed that

> The advantages proposed by [the graphical] mode of representation, are to facilitate the attainment of information, and aid the memory in retaining it: which two points form the principal business in what we call learning. . . . Of all the senses, the eye gives the liveliest and most accurate idea of whatever is susceptible of being represented to it; and when proportion between different quantities is the object, then the eye has an incalculable superiority. (*Breviary*, 1801, 14)

He understood that

> the eye is the best judge of proportion, being able to estimate it with
> more quickness and accuracy than any other of our organs . . . this
> mode of representation . . . gives a simple, accurate, and permanent
> idea, by giving form and shape to a number of separate ideas, which
> are otherwise abstract and unconnected. (*Atlas*, 1801, x)

He anticipated modern ideas in cognitive psychology such as depth
of processing by noting that people remember information better when
they process it in a meaningful rather than superficial way: "Information that is imperfectly acquired, is generally as imperfectly retained"
(*Atlas*, 1786, 3). He speculated that his charts were an aid to meaningful
memorization:

> Whatever presents itself quickly and clearly to the mind, sets it
> to work, to reason, and think; whereas it often happens, that in
> learning a number of detached facts, the mind is merely passive, and
> makes no effort further than an attempt to retain such knowledge.
> (*Breviary*, 1801, 6–7)

Playfair's circle and pie diagrams were intended to facilitate the comparison of land areas; comparing the irregular shapes formed by the
boundaries of countries in a conventional atlas was problematic and or-
dering countries by size was a difficult visual task. Playfair's solution was
to use a common shape and thus exploit the eye's capability of making
comparative judgments with high accuracy; "for where the forms are not
similar, the eye cannot compare them easily nor accurately" (*Breviary*,
1801, 15). Playfair was able to offer remarkable insights into the cognitive science of graphical perception two centuries before the flowering
of modern cognitive neuroscience.

PLAYFAIR'S LEGACY

Beniger & Robyn (1978) observed that, beginning with maps of Northern Mesopotamia, there was a 3,000-year-old tradition of representing
physical space (the world) by space (the map). Although sufficiently
inspired by mapmakers to use the word "atlas" in the title of his treatise, Playfair ended their monopoly on the use of spatial displays. His
genius was to realize that nonspatial quantities such as expenditures and
historical time could be represented by physical space and that such
representation offered advantages denied to tabular presentation. But

others did not share his conviction that he had found a superior way of presenting data, especially in his own country where concerns regarding accuracy were heightened by Playfair's carelessness, brashness, and disreputable personal reputation (Funkhouser & Walker, 1935; Funkhouser, 1937; Spence & Wainer, 1997). He was received more kindly on the continent where Humboldt thought highly of his creations (see Hankins, 1999), but there was still considerable opposition from many statisticians. Adoption of the new methods had to wait until the second half of the 19th century when Minard and Bertillon used some of Playfair's inventions in their cartographical work (Palsky, 1996; Friendly, 2002). In the United Kingdom, Playfair was almost completely forgotten until 1861, when William Stanley Jevons enthusiastically adopted Playfair's methods in his own economic atlas (noted by Keynes, 1936). Jevons (1886) wrote, "in statistics, the [graphical] method, never much used, has fallen almost entirely into disuse. It ought, I consider, to be almost as much used as *maps* are used in geography" (emphasis in original). Ironically, Jevons never succeeded in publishing his economic atlas, styled in the fashion of Playfair. Nonetheless, Jevons's advocacy of the graphical method found sympathetic ears in the British statistical establishment. Most influential among those influenced by Jevons was Karl Pearson, who not only embraced graphs, but included a lecture on charting in his famous series of statistical lectures at University College, London. Pearson acknowledged Playfair's contributions in generous terms.

In the 20th century, the use of graphs increased markedly and textbooks soon appeared. Brinton (1914) may have been the first widely sold primer on statistical graphs, but his text was quickly followed by a host of imitators. Today, it is normal to find graphs in newspapers, magazines, periodicals, and professional journals to communicate quantitative phenomena; and other visual media, such as television, make widespread use of charts for the same reason. Graphs are also used to explore and analyze data: Statistical charting is an integral part of almost all computer software used in the sciences and commerce. Playfair was well aware that charts were not merely a new way of presenting data to others. He recognized that graphs could stimulate new ideas or suggest models. After making a trial chart of some data, he said that "the first rough draft [gave] me a better comprehension of the subject, than all that I had learnt from occasional reading, for half my lifetime" (Playfair, 1805, xv–xvi). Playfair charted data to discover as well as to present; in that respect, he anticipated the exploratory uses of graphs that were to become popular at the beginning of the 20th century

(Spence & Garrison, 1993). William Playfair's vision, which he was unable to communicate to others during his life, affects and benefits us all. If he could see how his inventions have changed the ways in which we analyze and present data, he would be enormously proud.

Ian Spence
Howard Wainer

BIBLIOGRAPHY: VARIATIONS OF THE ATLAS AND BREVIARY

THE COMMERCIAL AND POLITICAL ATLAS

PRIVATELY CIRCULATED PRELIMINARY EDITION

Playfair, William (1785). *The commercial and political atlas: Representing, by means of stained copper-plate charts, the exports, imports, and general trade of England, at a single view.* London.

FIRST EDITION

Playfair, William (1786). *The commercial and political atlas; representing, by means of stained copper-plate charts, the exports, imports, and general trade of England, at a single view. To which are added, Charts of the revenue and debts of Ireland, done in the same manner by James Corry.* London: Debrett; Robinson; and Sewell.

SECOND EDITION

Playfair, William (1787). *The commercial, political, and parliamentary atlas, which represents at a single view, by means of copper plate charts, the most important public accounts of revenues, expenditures, debts, and commerce of England. To which are added charts of the revenues and debts of Ireland, done in the same manner, by James Corry, Esq.* London: Stockdale.

TRANSLATION OF THE SECOND EDITION

Playfair, William (1789). *Tableaux d'arithmétique linéaire, du commerce, des finances, et de la dette nationale de l'Angleterre.* Translated by Hendrik Jansen. Paris: Chez Barrois l'aîné.

LINEAL ARITHMETIC

Playfair, William (1798). *Lineal arithmetic: Applied to shew the progress of the commerce and revenue of England during the present century.* London.

THIRD EDITION

Playfair, William (1801). *The commercial and political atlas, representing, by means of stained copper-plate charts, the progress of the commerce, revenues, expenditure, and debts of England, during the whole of the eighteenth century.* London: Wallis.

THE STATISTICAL BREVIARY

Playfair, William (1801). *The statistical breviary: Shewing, on a principle entirely new, the resources of every state and kingdom in Europe; illustrated with stained copper-plate-charts . . . to which is added, a similar exhibition of the ruling powers of Hindoostan.* London: Wallis.

Playfair, William (1802). *Élémens de statistique: où l'on démontre, d'après un principe entièrement neuf, les ressources de chaque royaume, état et république de l'Europe: suivis d'un état sommaire des principales puissances et colonies de l'Indostan. Orné de cartes coloriées, représentant, d'un coup-d'oeil, les forces physiques de toutes les nations Européennes.* Translated from the English by Denis Francois Donnant. Paris: Batilliot et Genets.

REFERENCES

Author unknown (1823). Mr. William Playfair. *Edinburgh Annual Register, 16,* 332–4.

Belote, T.T. (1907). *The Scioto speculation and the French settlement at Gallipolis.* New York: Burt Franklin

Beniger, J.R., & Robyn, D.L. (1978). Quantitative graphics in statistics: A brief history. *The American Statistician, 32,* 1–11.

Biderman, A.D. (1981). The graph as victim of adverse discrimination and segregation: Comment occasioned by the first issue of information design journal. *Information Design Journal, 1,* 232–41.

Biderman, A.D. (1990). The Playfair enigma: The development of the schematic representation of statistics. *Information Design Journal, 6,* 3–25.

Brinton, W.C. (1914). *Graphic methods for presenting facts.* New York: The Engineering Magazine Company. Reprinted New York: Arno Press, 1980.

Broadie, A. (ed.) (2003). *The Cambridge companion to the Scottish enlightenment.* Cambridge, UK: Cambridge University Press.

Buchan, J. (2003). *Capital of the mind: How Edinburgh changed the world.* Edinburgh: Murray.

Cleveland, W.S. (1985). *The elements of graphing data.* Monterey, CA: Wadsworth.

Costigan-Eaves, P., & Macdonald-Ross, M. (1990). William Playfair (1759–1823). *Statistical Science, 5,* 318–26.

'Espinasse, M. (1962). *Robert Hooke.* Berkeley, CA: University of California Press.

Friendly, M. (2002). Visions and re-visions of Charles Joseph Minard. *Journal of Educational and Behavioral Statistics, 27,* 31–5.

Funkhouser, H.G. (1937). Historical development of the graphical representation of statistical data. *Osiris, 3,* 269–404.

Funkhouser, H.G., & Walker, H.M. (1935). Playfair and his charts. *Economic History (A supplement to the Economic Journal), 3,* 103–9.

Hankins, T.L. (1999). Blood, dirt, and nomograms: A particular history of graphs. *Isis, 90,* 50–80.

Herman, A. (2001). *How the Scots invented the modern world: The true story of how western Europe's poorest nation created our world and everything in it.* New York: Crown.

Hooke, R. (1705). Of the true method of building a solid philosophy, or of a philosophical algebra. In Richard Waller (ed.), *Posthumous works, containing his Cutlerian lectures, and other discourses, read at the meetings of the illustrious Royal Society.* London.

Jefferson, T., & Peterson, M.D. (ed.) (1984). *Writings: Autobiography / Notes on the State of Virginia / Public and Private Papers / Addresses / Letters (Library of America).* New York: Library of America.

Jevons, H.A. (ed.) (1886). *Letters and journal of W. Stanley Jevons*. London: Macmillan.

Keynes, J.M. (1936). William Stanley Jevons 1835–1882: A centenary allocation on his life and work as a statistician. *Journal of the Royal Statistical Society, 99*, 516–55.

Kosslyn, S.M. (1994). *Elements of graph design*. New York: Freeman.

Palsky, G. (1996). *Des chiffres et des cartes, naissance et développement de la cartographie quantitative française au XIXe siecle*. Paris: Comité des travaux historiques et scientifiques.

Playfair, J. (1822). *The works of John Playfair: With a memoir of the author*. James G. Playfair (ed.). Edinburgh: Constable.

Playfair, W. (1785). *The increase of manufactures, commerce, and finance, with the extension of civil liberty, proposed in regulations for the interest of money*. London.

Playfair, W. (1805). *An inquiry into the permanent causes of the decline and fall of powerful and wealthy nations*. London: Greenland & Norris.

Playfair, W. (1809–11). *British family antiquity illustrative of the origin and progress of the rank, honours, and personal merit, of the nobility of the United Kingdom*. London: Reynolds & W. Playfair.

Playfair, W. (1821). *A letter on our agricultural distresses, their causes and remedies: Accompanied with tables and copper-plate charts, shewing and comparing the prices of wheat, bread, and labour, from 1565 to 1821*. London: Sams.

Playfair, W. (1822). *A letter on our agricultural distresses, their causes and remedies: Accompanied with tables and copper-plate charts, shewing and comparing the prices of wheat, bread, and labour, from 1565 to 1821*, 3rd ed., with an additional chart. London: Sams.

Playfair, W. (1822–3). Unpublished ms, held by John Lawrence Playfair, Toronto, Canada. Transcribed and annotated by Ian Spence.

Priestley, J. (1765). *A Chart of Biography*. London: William Eyres.

Reid, T. (1764). *An inquiry into the human mind, on the principles of common sense*. Edinburgh: Kincaid & Bell.

Schofield, R.E. (1963). *The Lunar Society of Birmingham: A social history of provincial science and industry in eighteenth-century England*. Oxford: Clarendon Press.

Smith, A. (1805). *An inquiry into the nature and causes of the wealth of nations*, 11th ed., with notes, supplementary chapters, and a life of Dr. Smith, by William Playfair. London: Cadel and Davies.

Spence, I. (2000). The invention and use of statistical charts. *Journal de la Société Française de Statistique, 141*, 77–81.

Spence, I. (2004). William Playfair. *Oxford dictionary of national biography, Vol. 44*, 562–3. Oxford: Oxford University Press.

Spence, I. (2005). No humble pie: The origins and usage of a statistical chart. *Journal of Educational and Behavioral Statistics, 30* (3).

Spence, I., & Garrison, R.F. (1993). A remarkable scatterplot. *The American Statistician, 47*, 12–19.

Spence, I., & Lewandowsky, S. (1990). Graphical perception, Ch. 1., pp. 13–57, in J. Fox & J.S. Long (eds.), *Modern methods of data analysis*. Beverly Hills, CA: Sage.

Spence, I., & Wainer, H. (1997). William Playfair: A daring worthless fellow. *Chance, 10*, 31–4.

Spence, I., & Wainer, H. (2001). William Playfair (1759–1823): Inventor and ardent advocate of statistical graphics. In C.C. Heyde, & E. Seneta (eds.), *Statisticians of the Centuries*. New York: Springer-Verlag, 105–10.

Spence, I., & Wainer, H. (2004). William Playfair. In K. Kempf-Leonard (ed.), *Encyclopedia of Social Measurement*. San Diego: Academic Press, 71–9.

Tilling, L. (1975). Early experimental graphs. *British Journal for the History of Science, 8*, 193–213.

Tufte, E.R. (1983). *The visual display of quantitative information*. Cheshire, CT: Graphics Press.

Uglow, J. (2002). *The lunar men: Five friends whose curiosity changed the world*. London: Faber & Faber.

Valois, J.-P. (2000). L'approche graphique en analyse des données (avec discussions). *Journal de la Société Francaise de Statistique, 141*, 5–40.

Wainer, H. (2000). *Visual revelations: Graphical tales of fate and deception from Napoleon Bonaparte to Ross Perot*, 2nd ed. Hillsdale, NJ: Lawrence Erlbaum Associates.

Wainer, H. (2005). *Graphic discovery: A trout in the milk and other visual adventures*. Princeton, NJ: Princeton University Press.

Wainer, H., & Velleman, P.F. (2001). Statistical graphics: Mapping the pathways of science. *Annual Review of Psychology, 52*, 305–35.

Wainer, H., & Spence, I. (2005). Graphical presentation of longitudinal data. In B. Everitt & D.C. Howell (eds.), *Encyclopedia of Behavioral Statistics*. New York: Wiley.

THE
COMMERCIAL AND POLITICAL
ATLAS,

Reprefenting, by Means of

STAINED COPPER-PLATE CHARTS,

THE

PROGRESS OF THE COMMERCE, REVENUES, EXPENDITURE, AND DEBTS OF ENGLAND,

DURING THE WHOLE OF THE

EIGHTEENTH CENTURY.

THE THIRD EDITION,

Corrected and brought down to the End of laft Year.

By WILLIAM PLAYFAIR.

Printed by T. Burton, Little Queen-ftreet, Lincoln's-Inn Fields,

FOR J. WALLIS, NO. 46, PATERNOSTER-ROW; CARPENTER AND CO. BOND-
STREET; EGERTON, WHITEHALL; VERNOR AND HOOD, POULTRY;
BLACK AND PARRY, LEADENHALL-STREET.

1801.

PREFACE

THIRD EDITION.

I Have chofen the prefent moment for this new edition, not only becaufe the fubject is one of thofe that requires to be brought up, from time to time, but principally on account of the fingularity of the fituation in which Europe is now placed.

A great change is now operating in Europe, and though it is impoffible to guefs in what it will moft likely terminate, yet it is very certain that it will neither in a political nor a moral view

A 2   return

94260

return to its former fituation. The
minds of men, the boundaries of nations,
their laws and relations with each other,
are all in a ftate of change, and commerce muft feel the confequences of
thofe events of which it has been a principal caufe.

Should thofe revolutions and partitions
already effected, or about to be attempted, produce, as ufual, political fermenta-
tion in proportion to their importance,
Europe may probably be convulfed with
war for fifty years to come. The laft
century has been the century of arts and
commerce, this newly commenced may
then be that of war and contention. If it
turns out fo, a picture of the paft will
be a valuable thing, if, on the contrary,
commerce fhould ftill continue its progrefs, this will make the firft part of a
great whole, which, when completed on
fome future day, will be a moft valuable
work.

Had

Had our anceſtors repreſented the gradual increaſe of their commerce and expenditure, if it had not been an object of utility, it would at leaſt have been one of curioſity; but had records, written in this ſort of ſhape, and ſpeaking a language that all the world underſtands, exiſted at this day, of the commerce and revenue of ancient nations, what a real acquiſition would it not have been to our ſtock of knowledge? In place of which, a few detached facts are collected and brought forward as the only criterion from which we can judge of the manners and wealth of the ancient world.

It is not only of importance that this ſpecies of information ſhould be handed down, but alſo that it ſhould go down in ſuch a form and manner as that any perſon might even, though a native of another country, underſtand the nature of the buſineſs delineated.

Whatever

Whatever is rare or uncommon is an object of curiofity, and excites intereft; but when the fame object happens to be in itfelf of importance, the motives for inquiry become great beyond expreffion. We are gratified to know, that, in the times of Roman opulence, fifty-fix pounds was given for a lamprey, and that Lucullus expended fifteen hundred pounds for his fupper upon ordinary occafions, and certainly fuch facts are well worth knowing; but if we could have a copy of the cuftom-houfe books of Carthage or Tyre for a hundred years, what value might not be fet on them? Thefe charts will be for future nations the fame thing that the ancient records we fo much defire would be for us now, exhibiting the moft extenfive mercantile tranfactions that ever took place in the world, in a manner the moft fimple, eafy, and comprehenfive.

INTRO-

INTRODUCTION.

As knowledge increases amongst mankind, and transactions multiply, it becomes more and more desirable to *abbreviate* and *facilitate* the modes of conveying information from one person to another, and from one individual to the many.

Algebra has abbreviated arithmetical calculations ; logarithmic tables have shortened and simplified questions in geometry. The studies of history, genealogy, and chronology have been much improved by copper-plate charts. It is now sixteen years since I first thought of applying lines to subjects of Finance.*

At

* The Political Herald, (conducted by Dr. Gilbert Stuart, a man well remembered for his elegant literary talents) spoke of it thus at the time :—" The new method in which accounts

" are

At the time when this invention made its firſt appearance it was much approved of in England; Mr. Corry applied the ſame mode to the Finances of Ireland; and my original work was tranſlated and publiſhed in France, in the year 1788, when it was well received.

I confeſs I was long anxious to find out, whether I was actually the firſt who applied the principles of geometry to matters of Finance,

" are ſtated in this work, has attracted very general notice. " The propriety and expediency of all men, who have any " intereſt in the nation, being acquainted with the general " outlines, and the great facts relating to our commerce are " unqueſtionable; and this is the moſt commodious, as well " as accurate mode of effecting this object, that has hitherto " been thought of,

" Very confiderable applauſe is certainly due to this in- " vention, as a new, direct, and eaſy mode of conveying in- " formation to ſtateſmen and to merchants; although we " would recommend to the author to do whatever he can, " in any future editions, to make his leading ideas as familiar " as poſſible to every imagination, by additional illuſtrations " and directions; for theſe in ſome inſtances, ſeem to be " wanting." See vol. iii. pages 299, 305.

This laſt ſtricture is certainly juſt; and I have attended to the hint.

as

as it had long before been applied to chronology with great fuccefs. I am now fatisfied, upon due inquiry, that I was the firft; for during fifteen years I have not been able to learn that any thing of a fimilar nature had ever before been produced.

To thofe who have ftudied geography, or any branch of mathematics, thefe charts will be perfectly intelligible.* To fuch, however, as have not, a fhort explanation may be neceffary.

The advantage propofed, by this method, is not that of giving a more accurate ftatement

---

* When I went to France, 1787, I found feveral copies there, and, amongft others, one which had been fent by an Englifh nobleman to the Monfieur de Vergennes, which copy he prefented to the king, who, being well acquainted with the ftudy of geography, underftood it readily, and ex-preffed great fatisfaction. This circumftance was of fervice to me, when I afterwards folicited an exclufive privilege for a certain manufactory, which I obtained. The work was tranflated into French, and the Academy of Sciences, (to which I was introduced by Monf. Vandermond,) teftified its approbation of this application of geometry to accounts, and gave me a general invitation to attend its fittings in the Louvre; and at the fame time did me the honour of feating me by the prefident during that fitting.

than

than by figures, but it is to give a more fimple and permanent idea of the gradual progrefs and comparative amounts, at different periods, by prefenting to the eye a figure, the proportions of which correfpond with the amount of the fums intended to be expreffed.

As the eye is the beft judge of proportion, being able to eftimate it with more quicknefs and accuracy than any other of our organs, it follows, that wherever *relative quantities* are in queftion, a gradual increafe or decreafe of any revenue, receipt, or expenditure, of money, or other value, is to be ftated, this mode of reprefenting it is peculiarly applicable; it gives a fimple, accurate, and permanent idea, by giving form and fhape to a number of feparate ideas, which are otherwife abftract and unconnected. In a numerical table there are as many diftinct ideas given, and to be remembered, as there are fums, the order and progreffion, therefore, of thofe fums are alfo to be recollected by *another* effort of memory, while this mode unites *proportion*, *progreffion*, and *quantity*, all under one fimple impreffion of vifion, and confequently one act of memory.

This

This method has ftruck feveral perfons as being fallacious, becaufe geometrical meafurement has not any relation to money or to time; yet here it is made to reprefent both. The moft familiar and fimple anfwer to this objection is by giving an example. Suppofe the money received by a man in trade wete all in guineas, and that every evening he made a fingle pile of all the guineas received during the day, each pile would reprefent a day, and its height would be proportioned to the receipts of that day; fo that by this plain operation, *time*, *proportion*, and *amount*, would all be phyfically combined.

Lineal arithmetic then, it may be averred, is nothing more than thofe piles of guineas reprefented on paper, and on a fmall fcale, in which an inch (fuppofe) reprefents the thicknefs of five millions of guineas, as in geography it does the breadth of a river, or any other extent of country.

My reafon for adopting this mode of ftating the prefent revenue of the nation is for the purpofe of comparing it with the paft, as alfo of comparing the progrefs of the revenues of the ftate with the progrefs of the influx of wealth from other countries, for it is not from the prefent

fent ftate of things, uncompared with the paft, that any conclufion can be drawn.

The human mind has been fo acted upon for a number of years paft, and the fame fubjects have been fo frequently brought forward, that it is neceffary to produce novelty, but above all to aim at facility, in communicating information ; for the defire of obtaining it has diminifhed in proportion as difguft and fatiety have encreafed.

That I have fucceeded in propofing and putting in practice a new and ufeful mode of ftating accounts, has been fo generally acknowledged, that it remains only for me to requeft that thofe who do not, at the firft fight, underftand the manner of infpecting the Charts, will read with attention the few lines of directions facing the firft Chart, after which they will find all the difficulty entirely vanifh, and as much information may be *obtained in five minutes as would require whole days to imprint on the memory, in a lafting manner, by a table of figures.*

As to the materials, they are taken from the accounts laid every year before the Houfe of Commons,

Commons, therefore may be depended upon as the beſt that are to be procured. In the firſt edition I find the following obſervations, which, as they are equally applicable ſtill, I therefore ſhall inſert them.

" The giving form and ſhape, to what other-
" wiſe would only have been an abſtract idea,
" has, in many caſes, been attended with much
" advantage ; it has often rendered eaſy and
" accurate a conception that was in itſelf
" imperfect, and acquired with difficulty.

" Figures and letters may expreſs with accu-
" racy, but they never can *repreſent* either
" number or ſpace. A map of the river
" Thames, or of a large town, expreſſed in
" figures, would give but a very imperfect
" notion of either, though they might be per-
" fectly exact in every dimenſion ; moſt men
" would prefer *repreſentations*, though very
" indifferent ones, to ſuch a mode of painting.

" In an affair of ſuch conſequence, as the
" actual trade of a country, it is of much impor-
" tance to render our conceptions as clear,
" diſtinct, and eaſily acquired, as poſſible.

b " Infor-

" Information that is imperfectly acquired is
" generally as imperfectly retained; and a man
" who has carefully inveſtigated a printed table
" finds, when done, that he has only a very
" faint and partial idea of what he has read,
" and that like a figure imprinted on ſand it is
" ſoon totally erafed and defaced.

" The amount of mercantile tranſactions in
" money, and of profit or loſs, are capable of
" being as eaſily reprefented in drawing, as
" any part of ſpace, or as the face of a country;
" though, till now, it has not been attempted.
" Upon that principle theſe Charts were con-
" ſtructed; and, while they give a ſimple and a
" diſtinct idea, they are as near accuracy as is in
" any way uſeful."

On inſpecting any one of theſe Charts attentively, a ſufficiently diſtinct impreſſion will be
made, to remain unimpaired for a time, and the
idea which does remain will be ſimple and complete, at once including the duration and the
amount. Men of high rank, or active buſineſs,
can only pay attention to general outlines; nor
is attention to particulars of uſe, any farther
than as they give a general information; it is

11   hoped,

hoped that, with the affiftance of thefe Charts, fuch information will be got, without the fatigue and trouble of ftudying the particulars of which it is compofed.

The divifions that pafs from right to left are one million of pounds each. The divifions that pafs from the top to the bottom are ten years each. The crooked lines of exports and imports are meafured off upon the upright lines, according to the contents of the Tables added at the end.

Suppofe you want the amount of exports in the year 1750.—Obferve where the line of exports paffes the line marked at the bottom 1750, and by looking on the right hand margin, you will find it 13,300,000. The line of imports that fame year paffes at 8,600,000; and the difference between thefe two, which is 4,700,000, is the balance that year in our favour. In the fame manner, the amount of exports, imports, and balance, for any other year, may be found upon any of the Charts; though a very little practice will enable one to tell by the eye near enough, without any more trouble. Obferving the general figure of the whole is a good way to get at a diftinct idea of the nature of the trade.

In the particular Charts, the divifions from right to left are only one hundred thoufand pounds each, though in the general trade they are each one million, and in that of the national debt they are ten millions.

REMARKS

REMARKS

AND

GENERAL OBSERVATIONS

ON

CHART I.

REPRESENTING THE

TRADE OF ENGLAND

TO AND FROM ALL PARTS.

IN the general chart of exports and imports, during the laſt century, we ſee very nearly the *real* amount of our commerce, and with great exactneſs its *proportional* increaſe, for thoſe errors which ariſe from the nature of buſineſs, as it is and always muſt be tranſacted at cuſtom-houſes and ſea-ports, naturally bear the ſame proportion to the buſineſs done at one time as they do at another.

From the beginning of the century till the year 1750, our exports regularly increaſed faſter than our imports, ſo that the balance in favour of this country was greater then than it had ever been before; but, from that time, though our commerce has upon the whole doubled in its amount, yet the balance in our favour is not equal to what it was then : this is a proof that luxury has greatly increaſed amongſt us; and, not only in-

creaſed,

creafed, but that it has done fo beyond even the proportion
of our extended commerce.

The trade of this country received a great blow in the
years 1771, 1772, by the failure of fome great merchantile
houfes, which had been carrying on extenfive fpeculations,
fupported on a circulation of paper, from which it was juft reco-
vering when a revolt in our American colonies reduced it to a
very low fituation, in fo much, that, in the year 1781, the balance, for the firft time during the century, was againft us;
but, with the war, that difadvantage difappeared, and the
commerce, though not always with a regular pace, has increafed more rapidly ever fince then than at any foregoing
period.

It is evident from this chart that the trade of this country
was almoft in its infancy at the beginning of the laft century;
and now great beyond example. We fhall farther on in this
work have an opportunity of feeing that public expenditure
has increafed nearly in the fame proportion. It is impoffible to behold this rapid progrefs without concluding that
it muft come in time to a point which it cannot pafs, as nothing is infinite; it is therefore of great importance to trace
and find out to what caufes we owe our commercial fuperiority, that we may endeavour to prolong it as much as poffible;
for, though it may be a queftion admitting of difcuffion,
whether wealth, and what is commonly called commercial
profperity, is any real advantage to a nation, there can be no
queftion that the lofs of it, after having once enjoyed its poffeffion, is a very fevere misfortune.

Let us look at nations that never were rich—all is well
enough with them; but, let us look at fuch as *have been*
wealthy,

wealthy, and nothing is fo gloomy, dejected, and to appearance fo irremediably loft. From the banks of the Euphrates to the borders of the Scheld, we can trace the fhifting progrefs of commerce; and where, in former times, it flourifhed, we are always the moft certain to find poverty and want; or, at leaft, to find a liftlefs inactivity, which will in time bring them on. From Babylon, where commercial wealth was accumulated at a very early epoch, to Bruges, in Flanders, where it was at a very late period, the courfe may be traced with great eafe, and the refult will be found uniform and unequivocal.

Three great caufes have hitherto fhifted the feats of commerce : the pride which riches and luxury bring along with them, the envy which wealth excites in neighbouring powers, and the changes in the modes of carrying on trade, owing to difcoveries in the arts and in geography, have altered the channels through which it flowed.

That the empire of Babylon fell from its great pride and luxury is not a matter of doubt; Tyre fell through the ambition of Alexander the Great; Carthage and Palmyra, owing to the envy their wealth excited in the Romans. So vanifhed the grandeur of the commercial cities of the ancient world.— A few wretched huts for peafants or fifhermen, interfperfed with fcattered fragments of ancient palaces, are all that now remain of their former grandeur and glory.

Alexandria, founded by that great conqueror who gave it his name, as a fit emporium for trade, flourifhed for feveral centuries; but, as it was the connecting link between the ancient commercial world and the prefent, its downfall was owning to a double caufe. Firft, its wealth ex-

B 2   cited

cited envy, and it was facked and nearly deftroyed by thofe
defperate banditti who overrun and ranfacked the whole of the
civilized world, after the fall of the Roman empire; but ftill
the excellency of its fituation fecured to Alexandria confiderable trade, till the mode of intercourfe between the eaftern
and weftern world was changed, by the difcovery of the Cape
of Good Hope. Then it was that Alexandria funk entirely, and
that Venice and Genoa, which, in point of time, flourifhed
firft amongft modern cities, diminifhed greatly in importance;
the difcovery of the needle, and many other improvements,
which facilitated navigation on the ocean, tended generally to
reduce the confequence of thofe places, fituated on the borders
of the Mediterranean fea, which had till then engroffed the
wealth and importance of the weftern world.

Antwerp and Bruges, which were the depots of the
northern part of Europe for Indian productions, loft their importance in proportion as Genoa and Venice declined, and as
Amfterdam rofe into confequence.

Spain and Portugal, though not owing the wealth and importance they at one time enjoyed to commerce, but to conqueft, were foon enfeebled by the effects of that wealth; and
thofe two nations, whofe power and ambition fcarcely knew
any bounds about two centuries ago, are now in a very humble fituation. Spain is entirely fubfervient to the will of
France, and Portugal is reduced to borrow from England a
paultry fum to enable it to protect itfelf againft the French,
or to purchafe an inglorious tranquillity.

Holland too, which was a commercial and rich country before England was fuch, has been for half a century on the decline;

cline; and, since it lost its independence, by the entry of the
French into it, can scarcely be considered as of much importance in the commercial world; however, though on the decline, and considerably advanced in it, matters may yet take
another turn, for pride and luxury are so powerfully counteracted by avarice in that country, that they will not produce such
baneful consequences as in those other parts of the world, of
which we have been taking a cursory view.

As we find then that the transitory prosperity which commercial wealth gives to a nation is succeeded by poverty and
insignificance, if not by slavery and wretchedness, it is of
great importance for us to endeavour to counteract this natural tendency, and to procure for this country a continuance of
those blessings which have hitherto enabled us to support
greater burthens than any nation ever before did.

Our great commerce and naval power certainly excited
the envy of all those who took part against us during the American war, but though our enemies succeeded in wresting our
extensive provinces from us, yet our national energy has prevented those bad effects that we feared and they expected, inso-
much, that our commerce is more extensive than ever; and if
we can be but *moderate*, *just*, and *prudent*, there is reason to hope
that we may at last fix the abode of commercial wealth and
prosperity in our island, not with any wish to engross it from
the rest of the world, but to preserve ourselves in that exalted situation which we at present enjoy.

Our manufactures are daily improving at home. The population of the country is as it were augmented by those
active but inanimate machines that perform the work of more
than three millions of people, without the expense or con-

B 3   sumption

fumption of food and clothes, and at the fame time confumers for our manufactures are daily increafing in number in America, which circumftance may give fupport to our manufactures for ages to come, as it will not foon be the intereft of the inhabitants of fo rich and extenfive a country to become manufacturers* themfelves.

The two years in which our commerce funk the moft rapidly, as appears by the chart, were 1772 and 1793, on which years there were great bankruptcies, and the American trade was that which fuffered the moft. The reafon of this is evident, and the conclufion to be derived important. The American trade depends more on credit than any other, and as failures do a general injury to the credit which is fupported by bills, and which enables manufacturers and merchants to give the long terms of payment that are granted to thofe who carry on bufinefs in America, it was naturally that branch of trade which was the moft affected.

It may not perhaps be improper here to obferve, that the great extent of our trade, though owing in part to the good quality of our manufactures, is ftill more favoured by the long credit we give to foreigners; for, thofe who keep ftore-houfes, fhops, or magazines abroad, can fill them with Englifh goods on credit, whereas they muft pay almoft ready money for commodities manufactured at home, they therefore not only find a facility in procuring Britifh merchandize, but they

* Every inhabitant of America may be confidered as confuming annually Englifh manufactures to the amount of a guinea, and as they are increafing at the rate of doubling in fifteen years, the market for Englifh goods is extending every day, and in thirty years America alone may be able to confume all the manufactures we can produce.

prefer

prefer felling, for ready money, what they can replace upon credit. From this it arifes that Englifh goods are in a manner forced upon the confumers, and that thofe who deal in them, being eager to fell, do it on a moderate profit. It is to a well-managed fyftem of paper credit that we owe the power of doing this, as the real monied capital of the nation is in a great meafure abforbed by the public funds.

Time

EXPORTS & IMPORTS
to and from
IRELAND.

Exports of England

BALANCE in FAVOUR of ENGLAND

Exports of Ireland

Imports from Ireland

Line of Imports

Line of Balance

Money

1700 1710 1720 1730 1740 1750 1760 1770 1780 1790 1800

100,000 L
2
3
4
5
6
7
8
9
1 Million
1.1
1.2
1.3
1.4
1.5
1.6
1.7
1.8
1.9
2 Millions
2.1
2.2
2.3
2.4
2.5
2.6
2.7
2.8
2.9

Neele sc. Strand.

TRADE with IRELAND.

CHART II.

FROM the beginning of the laſt century, till the year 1720, we find the trade with Ireland was very inconſiderable, and that the balance, though ſmall, was againſt England.

From that period, however, this branch of trade has increaſed with conſiderable rapidity and regularity, both in the whole amount and the balance, which ever ſince has been in favour of England.

Before the Iriſh nation was ambitious of rivalling England in manufactures and commerce, their capital and their labour were employed in thoſe things that were moſt naturally the productions of that country. Had their induſtry or wealth been too great to find employment on ſuch objects, any liberties tending to increaſe that field would have enriched the country. Unluckily for Ireland, however, the very contrary was the caſe : thoſe very liberties produced a different effect, and their views were diverted to objects which to them are much leſs advantageous ; for the manufactures that ſettle firſt and naturally in a country are, in general, the moſt advantageous, and thoſe eſtabliſhed by wealthy or ambitious individuals, in imitation or in rivalſhip to other nations, are generally ephemerical and often ruinous.

Manufactures

Manufactures are of all things the most difficult to tranf-
plant, becaufe the habits of the people muft change before
they can thrive,* and alfo becaufe they muft be lofing concerns, till they come to fuch perfection, as at leaft to equal
thofe of other countries. Nature has in general been fo careful to point out, by difference of climate, foil, or fituation,
the manufactures that fuit a country beft, that it is our fault
when we miftake her intention; and whenever recourfe muft
be had to prohibitions, premiums, and fuch things, to en-
courage common trade, there is reafon to fear that fome miftake has been committed. Thofe manufactures which thrive
beft with people of great capital are the very worft for
thofe who are not rich, as they bring on a very unequal fort
of a competition, which is generally hurtful to the poorer
party, though, when carried on with much moderation,
it fometimes ends differently; and the attentive induftry neceffary to the poorer defcription of men has occafionally, in
the end, triumphed over the capital of the rich and negligent.

Should a feparation of interefts ever take place between
England and Ireland, a circumftance now, not likely to hap-
pen, it will be to their mutual difadvantage. England will
lofe power, and Ireland will be ruined.

* That this is true, may eafily be feen in England, and in moft countries, as women and children affift in every confiderable local manufacture,
and muft be bred up to it. In countries that manufacture linen or woollen,
women and children univerfally card or fpin; and when once bred up to
that, they could never handle a hammer or a file with any degree of dexterity. Many women about Birmingham and Wolverhampton never faw
a fpinning-wheel; but they are very expert at making nails, buttons,
buckles, &c. &c. No manufacture where women and children do a great
part of the work can be tranfplanted in lefs than a lifetime, and feldom
then.

In

In any dispute between two countries so connected as England and Ireland, the richer nation has at first the disadvantage; it has nothing to get, and much to lose; the other, on the contrary, has much to expect, and is not afraid of losing any thing: these causes operate in producing presumption in the one, and timidity in the other. When, however, a disagreement has fairly come to a head, and force comes to be opposed to force, the matter is intirely altered; and though the interference of other powers may render unsuccessful the advantage of the richer country, they can neither alleviate nor remove those evils that attend the poorer, which do not end with the contest, but are continued and extended on account of the self-defence, and other expenses of government.

Were it not too presumptuous as well as tedious to investigate or decide upon a question so agitated, as that of the benefit which will be reaped by the Union with Ireland, there is ample room for observation. The interest of both countries required a change of system; and justice demanded, that it should be upon fair principles. It would be offering an insult to the judgement of the leading men on the different sides of the Channel to suppose, that they thought any good could possibly arise from a separation. The case has frequently been assimilated to that of North America, though it is very different, and resembles not, in any respect, a portion of that continent, which, extending almost from the torrid to the frigid zone, nature never could intend as an appendage to a distant, luxurious, and divided island.

Ireland, in case of a separation, must claim protection from some other power, and none, unequal to England by sea, could grant that effectually. For Ireland to maintain

a fleet

a fleet and an army in any degree *adequate* to maintain independence, it muſt be ſubjeſted to expenſes to which the nature of the country itſelf is very unequal; for that nation, though fertile and produſtive, is not peopled with a race of inhabitants who have the turn for what gives importance to Britain, manufaſtures and commerce.

It will readily be allowed, that neither population nor extent of territory give either England or Ireland much title to political importance, and as they united enjoy a great deal, it muſt be attributed to commerce and ſituation, in both of which Ireland has a manifeſt inferiority.

It is to be hoped, when the eyes of the inhabitants will be opened to thoſe advantages which are yet but in proſpeſt, and to which national pride prevents them at preſent from giving a full belief, that the unpleaſant feelings which any union of countries is generally calculated to produce, and perhaps too, when not carried into effeſt in a manner the moſt mild or agreeable, will be entirely done away. We view with a jaundiced eye whatever we deem compulſory, though ultimately favourable to ourſelves, or whatever tends to infringe what we may ſuppoſe the natural independence and rights of mankind.

TRADE

Plate 3

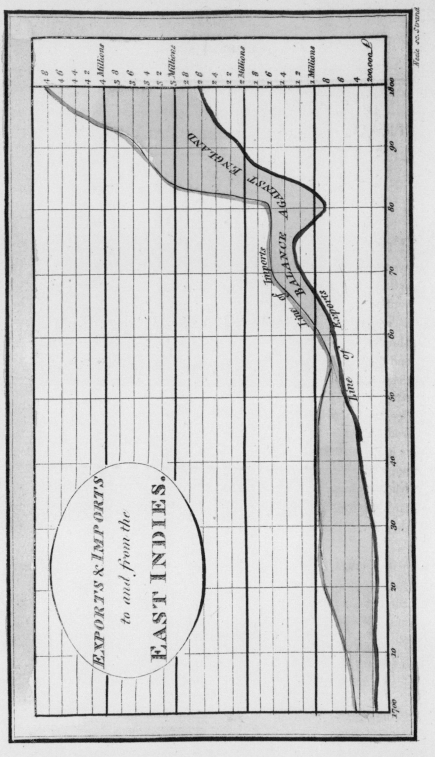

EXPORTS & IMPORTS
to and from the
EAST INDIES.

Line BALANCE AGAINST ENGLAND

Line of Imports

Line of Exports

4 8
4 6
4 4
4 2
4 Millions
3 8
3 6
3 4
3 2
3 Millions
2 8
2 6
2 4
2 2
2 Millions
1 8
1 6
1 4
1 2
1 Millions
8
6
4
200,000 £

1700
10
20
30
40
50
60
70
80
90
1800

Neele Sc. Strand

TRADE to and from the EAST INDIES.

CHART III.

THE following question will immediately suggest itself to every one : does this small chart, does the insignificant sum here delineated, represent truly the transactions of a company, whose confidential servants are princes, whose inferior officers rival in wealth the richest nobility, and whose meaner servants plunder, with impunity, the native inhabitants of the richest and finest portion of the world? The disproportion between the wealth that is acquired by their servants, and by the company, is very evident at the first view ; nor is the cause itself deeply concealed. All that portion of the riches that come from India, which is originally got by rapacity, must belong to individuals ; for, if men are to pass the line of equity, it will not be for the purpose of enriching a company of merchants, who have never braved the dangers of war, or encountered the far more dreadful ones of an unhealthful climate, but to serve themselves.

It does not appear that the affairs of India have ever been conducted upon a right principle. While we have been sovereigns of that country, and preserved it under a despotic military government, the desire of selling, exclusively, in Europe the productions of Asia, has made it so complicated a system, that it scarcely admits of any such thing as good management.

It

It is not to be expected, that men of all defcriptions, going to a diftant country, and a dangerous climate, will return, without endeavouring to amafs fortunes, as fome recompence for the lofs of health and conftitution. Were it poffible to prevent that accumulation, nobody would be found foolifh enough to go to India : nor is there any inducement, lefs than that of a large fortune, which could make men do, what, even with that advantage, it is frequently a misfortune to have done.

The idea, however, of preventing thefe various fcenes of extortion that are faid to be practifed upon the unhappy inhabitants of that rich country, was an honour to the human heart; I allude to the proceedings in Parliament prior and fubfequent to the trial of Mr. Haftings, but the manner in which it has been attempted has not done equal credit to the underftanding. To prevent *injuftice*, is it neceffary to be *unjuft* ? To prevent *oppreffion*, is it neceffary to be armed with the powers to *opprefs* ? Such, however, have been the modes propofed for the removal of grievances in India. During the exiftence of the Roman empire, there were many cafes parallel, in fome refpects, to that of which we now fpeak. The Roman governors generally plundered the provinces, and, returning to Rome, loaded with riches, frequently underwent examinations. They, too, had their *pains* and *penalties*, and *reftraining bills*, to prevent or to punifh this evil : and, as the features of Roman law were ftrong, as even their popular government was pretty arbitrary, they foon enacted very fevere laws againft this fpecies of oppreffion. The confequence generally was, that, before a governor dared to meet inveftigation, he not only brought home a fortune, but immenfe fums, to fecure, by bribery and corruption, that impunity he had not to expect from juftice. The provinces then groaned under a double load of oppreffion. This, it is probable,

bable, may be the cafe with India. Indeed, there can be little doubt that it *will* be the cafe : for, befide the example furnifhed by the Roman world, there is a fact that might be evident to any obferver, with regard to crimes and punifh- ments: that, when punifhments are very fevere, they increafe the enormity of the crimes they were intended to prevent; and they diminifh the number, *only* when there is a pofi- bility of keeping *altogether clear* of tranfgreffion.

As all people who go to India, even though they may be in other refpects honourable men, are under a neceffity of get- ting more money than the company allows, they mufi tranf- grefs that line, the paffing of which fubjects them to punifh- ment; and, when once paffed, the utmoft feverity can only act as it did with the Roman governors, by making rapacity ftill more rapacious.

Should it be propofed to put laws fo very rigoroufly in force, as to prevent all bribery, or corruption of every fort, then might it indeed prevent the evil of which we have been fpeaking; but then alfo another would arife. The rich ad- venturers in India, preferring the unwholefome climate in which they acquired their wealth, to the infolence of office, and the law's delay, to the horrors of a dungeon, and that which of all things is the moft unfupportable, infamy and dif- grace in their native country, would never return ; and India, after evading our efforts to govern it right, would govern it- felf; it would open its ports to the nations of the weft and of the north, and would become itfelf the richeft na- tion in the world. They have but unanimoufly to pafs the vote, and they are not a moment longer the fubjects of Great Britain. To what refources could we apply, to

wage

wage war upon an army of veteran Europeans, at the diftance of almoft half the globe, and poffeffing wealth, and property, and every means to defend themfelves againft us, for a longer period than we could perfift, even were we to mortgage our country?

The great wealth derived from this branch of commerce is brought over by individuals, and has occafioned an immenfe increafe of riches, luxury, and extravagance, in this country, and tends greatly to the precipitating us into that decline, that fooner or later overtakes all commerial nations. Nothing can be more hurtful to real induftry, than to perceive the fuccefs in acquiring wealth of thofe who come from that part of the world.

NOTE. The fimilitude between the affairs of the Roman empire and thofe of our India company (to compare great things with fmall) feems to have held alfo in the proportion of the wealth of individuals, and of the ftate; that of the former being very immenfe, compared with that of the latter.

Lollia Paulina, *the niece of a Roman governor*,* could afford to wear, in her common drefs, jewels to the value of 322,916*l.* 13*s.* 4*d.*

Pallas, a freedman of Claudius, † and keeper of his privy purfe, was reckoned worth 2,421,875*l.* which was all acquired in a fhort fpace of time. In general, the affairs of the wealthy people at Rome were upon this immenfe fcale. Pompey's falary,‡ during four years of his government, was yearly 193,750*l.* Yet, when Julius Cæfar pillaged the treafury, at the beginning of the civil war, he found in gold, filver, and in money, (for he took both bullion and money) only 1,095,979*l.*§ which was comparatively a very fmall fum. It was not even equal to the debts that Cæfar had contracted, without any other than perfonal fecurity; for, though Craffus, and other rich men, were bound in large fums for him, yet it was only after his creditors would not let him go to his province.

* Plin. Lib. 9, Cap 15. † Tacit. Lib. 12

‡ Plutar. in Pompeio. § Plin. Lib. 3. Cap 3.

This

This is a clear proof of the fmallnefs of the wealth of the ſtate, when compared with that of individuals, who had provinces to pillage.   There is no reaſon to believe that at any time the revenues of that immenſe empire were above eighteen millions Engliſh money; a ſmall ſum, when compared with the extent of territory, and the manners of the times.   The concluſion is pretty fair, that thoſe maſters of the world, who had conquered, and actually got rent for great part of the lands in the empire, when they had taxed every thing that was taxable, even *ſmoke, air, and ſhade*,* did not receive great ſums into their public treaſury, but that the individual miniſters of their oppreſſion, like thoſe of later times, retained in their own poſſeſſion the far greater portion of the wealth extorted from the provinces.

* Zonaras. fumum, aërem, et umbram.

TRADE

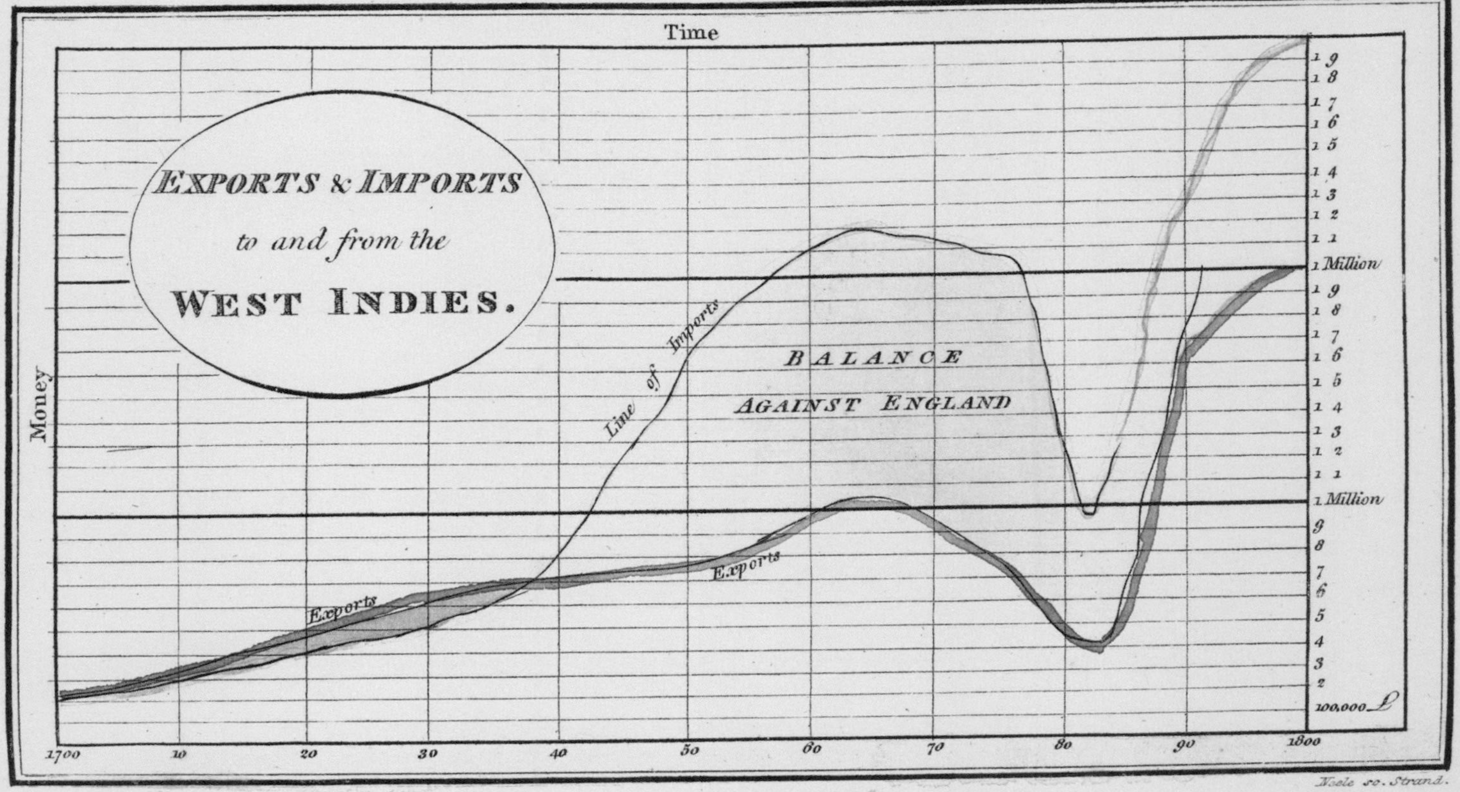

Plate 4.

£ 100,0000

Scale 40. Strand.

EXPORTS & IMPORTS to and from the WEST INDIES.

Time

Money

BALANCE AGAINST ENGLAND

Line of Imports

Exports

Exports

1 Million
1 .9
1 .8
1 .7
1 .6
1 .5
1 .4
1 .3
1 .2
1 .1
1 Million
.9
.8
.7
.6
.5
.4
.3
.2
.1

1700 10 20 30 40 50 60 70 80 90 1800

TRADE TO AND FROM

THE

WEST INDIES.

CHART IV.

THE poffeffion of the Weft-India Iflands has proved of much advantage to Britain; and the balance in this trade is of a nature totally different from that with any independent country. Though apparently againft us, it is really in our favour. Did thefe iflands import as much as they export, the poffeffion of them would not be of much value. Of the articles which we import from thence, part is again exported from this country to the continent. The planters alfo, and other fettlers, who generally return home, bring their wealth with them: fo that in its nature it is very different from the other branches of trade, and even from that which was carried on with the Americans while they fubmitted to our government. The trade with America was the fame in its nature with any foreign trade, and the balance that fairly appeared in our favour and no more was real gain. Every fettler in America became an American; every fettler in the iflands almoft continues to remain an Englifhman; fo that, with Jamaica and the other iflands, we may be faid to be carrying on an internal and not a foreign trade. With America it was entirely foreign. There is a radical defect in the Englifh fyftem with the colonies or poffeffions abroad. The riches from which, come into the pockets of a number of individuals, and the expenfes of which, fall upon the nation at

c 2   large.

large. This will in time multiply taxes, and facrifice general good to a particular object and to private advantage.

Thefe poffeffions, as defirable for the wealth they afford,
as infamous for being of the number of thofe fpots where
European avarice triumphs over all the virtues of humanity,
afford us rum and fugar, at the expenfe of the lives and freedom of the much injured, and wretched inhabitants of Africa.
But it is to be hoped that this abufe will find a remedy; as
much attention has been lately paid to the fubject, and moft
well informed men are of opinion that a change might eafily
be effected that would be productive of general advantage.

OBSER-

Plate 5

Time

Money

EXPORTS & IMPORTS
to and from all
NORTH-AMERICA.

BALANCE in FAVOUR of ENGLAND

BALANCE in FAVOUR of ENGLAND

Line of Exports

Imports

Line of Imports

Exports

1700   10   20   30   40   50   60   70   80   90   1800

5.8
5.6
5.4
5.2
5 Millions
4.8
4.6
4.4
4.2
4 Millions
3.8
3.6
3.4
3.2
3 Millions
2.8
2.6
2.4
2.2
2 Millions
1.8
1.6
1.4
1.2
1 Million
8
6
4
200000

OBSERVATIONS on the TRADE

WITH

NORTH AMERICA.

CHART V.

PERHAPS no kingdom ever formed a more great or noble scheme, than that of peopling, governing, and protecting an eighth part of the known world; but the thing was in itself impracticable, it was too great a project, and its principles were unsound. We expected that obedience from a child, that has only sometimes been exacted from a slave.

There are particular spots on the earth that are rich by nature, and seem to court the yoke from the inhabitants of poorer countries. Such are the Spice-islands, and other possessions in the East Indies; the West-India islands, and some parts of South America; great riches are sometimes derived from extending dominion over such, and importing their produce. The Romans, indeed, acquired riches by dominion over *poor nations*, but then it was by a *tributary revenue*, it was not a *commercial one*, neither were the nations founded and nursed at their expense.

The British empire followed a different plan from either of these, in peopling America. It was at the expense of peopling, protecting, and governing a distant country, the situation, extent, and nature of which were such as insured it independence, whenever it should think proper to make the demand.

Things

Things took their natural courfe, and America, always a very expenfive poffeffion, at laft afferted its own liberty, and was fuccefsful, but notwithftanding the failure of the projeĉt of preferving it as a Britifh colony, the advantage from American trade is greater than ever to England.

Hiftory, perhaps, does not furnifh a greater inftance of the downfal of ambition, and the vanity of human projeĉt, than Britain experienced in the revolt of America. The idea of preferving it in a ftate of dependence was rapacious, impolitic, and unjuft. Happy would it have been for both nations had it then been confidered fo by the mother country. England might have faid, ' We never expeĉted to derive any
' other revenue from you than what may be the confequence
' of a mutual trade ; nor were we ever foolifh enough to con-
' ceive that you would trade with us but when it was your
' intereft ; the extent of your coafts, and your diftant and
' continental fituation, prevent that. You are able, and you
' wifh to be independent ; let us part friends, and deal as ex-
' tenfively as it is our mutual intereft to do ; that is all we can
' expeĉt of each other.' The Americans will probably not attempt thofe manufaĉtures with which they have hitherto, and may ftill be fupplied from England. It will not be their intereft to do fo for many years yet to come ; for though it is not impoffible that they may attempt it, it is very improbable that they will perfevere. The fame divifion of labour that takes place in fingle manufaĉtories, takes place alfo in towns, counties, and nations ; and the advantages arifing from it muft have been very foon perceived ; for all nations are acquainted with barter and exchange ; and without divifion of labour there would have been very little of that in the world.

The

The manner in which commerce is now carried on to America is very different from what it was at the beginning of the laſt century, or from what it had been in any former period, or to any other country.

In the infancy of commerce, a demand for commodities preceded the manufacture or the traffic in them ; and before a ſhip was freighted to carry goods to a diſtant port, there was ſome degree of certainty that they were wanted at that port ; nor did the inquiry ſtop at whether they could uſe the things, but whether alſo they would be willing to pur-chaſe them, and pay the money. Speculations then were limited, and every thing was reduced to a certainty, except the dangers of the ſea. It is different now : the dangers of the ſea are reduced to a certainty ; but whether the goods are wanted, or will be paid for, is often very uncertain. This has been in a peculiar manner the caſe with our commerce to America. Though every ſtate of things does, during its exiſtence, naturally undergo changes and refinements that were unknown at firſt, and mercantile buſineſs has, like other things, improved in its modes, and though letters and foreign correſpondences have long ſuperſeded the neceſſity of merchants going to ſea in perſon, yet nothing can ſuperſede the neceſſity of the purchaſers of the goods being in a con-dition to pay for them. This, however, was not thought worth while to inquire into, with our own countrymen and our children, as we were pleaſed to ſtyle them in America. So much was this the caſe, that if a young man, who was known to have been in the ſervice of a merchant, and not entirely deſtitute of ſenſe or conduct, choſe to apply for credit in England, upon the faith of having opened a connec-tion in America, he could get ten times more credit than a

ſober

fober, induſtrious man, who confined his trade to his own
country.

Until about the middle of the laſt century, the North
Americans, who have neither gold nor ſilver produced in their
country, and who therefore muſt pay in merchandiſe and not
in money, uſed to ſend over produce to a greater amount than
they bought of our manufactures; we were rich, and could
pay a balance in hard caſh, the thing that they ſtood in need
of. This was right; it was juſt as it ſhould be, and as it
would have continued to be, had not a change taken place in
this country, which put an end to it, and the natural proportion was loſt, in an inſtant, between what they bought
and what they ſold: that regular proportion then, which
half a century had ſeen take place, was at once entirely done
away. It was about the ſame time that a change took place
among our merchants; and a number of men aſpired to that
lucrative and reſpectable ſituation, who had no property to
riſk, no money to loſe, and who were willing to play with
the property of other men, to their own advantage. America,
a branch of our empire, the ſame language, people, and religion, afforded a wide field, and it was occupied.

Did a ſpeculator, who went in this looſe manner to work,
prove ſucceſsful, by getting returns, he became a rich Engliſh-
man; and if he failed, America, the place to which he had
conſigned the fortunes of his creditors, afforded an aſylum from
all the local inconveniences of his misfortunes.

In this very manner, (and it muſt be remembered by many
people now living,) did the American trade increaſe: falſe
capitals and falſe credits increaſed alſo; they exiſted depen-
                                                            dent

dent upon each other; and as a proof of this, in the year 1771-2, which will long continue to be remembered as the time when perfonal credit received an uncommonly fevere check, we find this trade decreafe no lefs than two millions. Other branches of trade did not feel this fhock in any fimilar degree.   It is therefore fair to conclude, that between falfe credit and American trade, there exifted fome particular connection.   Had not the war or fome other circumftance happened to put an end to this exportation bufinefs, the capital of our Englifh merchants, or rather of the Englifh manufacturers, might have continued to march over to America, and we fhould imperceptibly have received an injury, of the extent of which we were not aware.

Were this not a thing that has fo lately happened, that moft people muft remember, and fome people feel the effects of circumftances that confirm it, fome time might be well employed in bringing proofs; of which, as matters are, a few may ferve.

Moft part of thofe young adventurers, who have begun without capital of their own, or knowledge, have begun in the American trade.

Englifh manufactures are faid to have fold frequently, even fince the war, as cheap, and fometimes cheaper in America than they have done in England.

Could thefe things poffibly have happened, if American trade had not been carried on under different rules and aufpices from other branches of commerce, or from what it was in the former part of this period?

There

There feem to have been three caufes for the rapid increafe of American imports:—The actual wealth and capital
of this country, paper credit, and the increafed population of
America.

Let us firft confider the wants of men, and their abilities to
fupply them. Do they not go on in a neceffary proportion, and
does not the ability generally precede the fupply? As we poffeffed almoft all the trade of America, its increafe admits of
more accurate reafoning than any of the other branches, of
which we only poffeffed a fmall portion, which portion might
therefore increafe or decreafe, without the whole amount of
the trade undergoing any material alteration. ·

The value of the goods that we imported from America is
probably a pretty fair meafure of their abilities, and fhews the
rate at which their wealth increafed.   If their riches increafed
in that line, then fhould our exports to America have rifen
nearly at the fame rate: and accordingly, till the year 1755,
it goes in a direction nearly parallel; but after that, it goes in
a line fo entirely different, that there cannot remain a doubt,
that to produce fo very different an effect, another caufe muft
have begun to operate.   Nor is there any caufe fufficiently
great to have produced this effect, or likely to have done it,
except paper credit which we have already been confidering.

Ever fince the invention of paper credit, trade has had a
latitude it did not before enjoy, and its progrefs being lefs
natural, has become more intricate.   That bound fet and preferved by the nature of things was removed, when paper credit was firft invented; previous to which, nothing reprefented wealth that was not wealth itfelf, or that was not phyfically worth the fum that it reprefented; and in order to give
credit

credit in bufinefs, it was abfolutely neceffary either to poffefs, or to have borrowed a capital. Nations could not then extend their trade; the trade extended itfelf, and, like other natural productions, grew in proportion to what it had been. The effects of this invention it is not neceffary to inveftigate in every point of view. It has been of infinite utility to mankind upon the whole; at the fame time that it was undoubtedly the means of enabling this falfe ftructure of American trade to be raifed with the greater conveniency. By means of paper the inconveniency of giving long credit is in a great degree taken away; and very long credits were neceffary in trading to America. Befides that, as drawing bills produces, in the firft inftance, the fame effect with having difcovered a treafure, though in the end often operating as if one had been loft, it leads to the fpeculating too far, and being involved before it is perceived by the individual that he is in any danger. This caufe alone, however, could not have operated fufficiently to produce fo great a trade, had not the other of the actual wealth of our manufacturers enabled them to produce the goods, and the increafing population of America procured confumption.

Upon the whole, this chart exhibits, under different circumftances, a very ftrong and a very curious contraft.

For the firft fifty years, we obferve the fimple and regular growth, from poverty to wealth, of a new country; during the fucceeding twenty years, we are aftonifhed at the extent and operation of a mad mercantile fpeculation carried on by our own country; and the period which fucceds, fhews the cataftrophe that fo airy and fo ill-founded a project was likely, fooner or later, to experience. There is not any branch of trade, which, from the nature of its progrefs, affords fo much in-

ftruction

ſtruction as this. It merits equally the attention of the philo-
ſopher, the politician, and the merchant; for it throws light
upon all the three different objects of their purſuits.

Upon the manner in which buſineſs is conducted, depends
ſomething more than merely the gaining or loſing a little
money. The happineſs of numbers of innocent individuals is
frequently depending upon the ſucceſs of projects, with the
formation of which they had no concern. What numbers have
been ruined, and how many more have been deprived of fortune, by our ill-conducted trade with America?

It is an improper ſubject for diſcuſſion here, whether happineſs is or is not of more importance than exiſtence itſelf:
many people think with Julius Cæſar that it is; but we run
no riſk certainly in affirming that it is of *very great* conſequence; and that, as it in a conſiderable degree depends upon
the ſucceſs of mercantile affairs, they, as well as the methods
of prolonging life or procuring health, deſerve our attention.

To acquire a knowledge of the nature and arrangement of the
human frame, ſeminaries are erected, and the graves are ran-
ſacked; from the conviction, that unleſs it is made a particular ſtudy, and unleſs former caſes are attended to and under-
ſtood, little can be done in preventing future diſeaſe. No
pains or attention is taken, however, to prevent thoſe evils
that wrong projects and unſucceſsful mercantile ſpeculation ſo
widely diffuſe. This has not, except by a few individuals,
ever been conſidered as an object deſerving attention. Nor
has, in this ſtudy, any great attention been paid to the connection between cauſe and effect; a connection that it is neceſſary to know, and to underſtand which, the compariſon
between paſt events is indiſputably the moſt obvious, as well

9   as

as the moſt certain road. Men in general are very ſlow to enter into what is reckoned a new thing; and there ſeems to be a very univerſal as well as great reluctance to undergo the drudgery of acquiring information that ſeems not to be *abſo-lutely* neceſſary. It ſeldom appears uſeful, in a high degree, to underſtand ſubjects that have not hitherto been objects of attention. The capital, the arts, and the induſtry, of this country, are too great to be entirely employed without ſome ſpeculation; the principles, therefore, on which ſpeculations ſhould be made, will become an object deſerving and requiring attention, not leſs ſo than the art of preſerving the health of the human body.

With all due regard to the opinion of mankind, which ulti-mately ſtamps a value, or its oppoſite, on human inventions, the author of this preſumes to think that the mode of painting to the eye the tranſactions of paſt times, is a conſiderable ſtep in making that inveſtigation eaſy, which he apprehends to be ſo neceſſary. It is in order to make uſe of viſion that the ana-tomiſt lectures over a dead body; the mathematician over a figure drawn upon a ſurface; and the aſtronomer over his globes and orreries: without which, the labour would be in-creaſed, and the progreſs diminiſhed. The lines here uſed to repreſent quantity and time, do it with a mathematical exact-neſs that is not to be called in queſtion, and enable the ſame advantage to be obtained that thoſe ſciences derive from viſion.

The hiſtory of the world has furniſhed few inſtances of ſo great a tract of country undergoing a change, from an uncul-tivated and barbarous, to a civilized ſtate; and it will well merit the attention of mankind to obſerve the different ſteps and the progreſs upon ſo large a ſcale.

There

There feems to be fome ftrange enlivening vapor in a new foil, that an old one does not poffefs; for the human mind is degenerated in thofe parts of the world which once gave laws to the reft, and has rifen vigorous and frefh, in places which were then infignificant. Britain, in the days of Auguftus, was peopled with painted barbarians; and, in later times, Ruffia exhibited the fame appearance; yet who fo active in improvement as they? In that ftate of men, when they endeavour to rival their neighbours, there feems a fort of enthufiafm that impels individuals, and invigorates the ftate. The nation that has already got to the defired rank, begins to relax the exertion, and to enjoy its fruits; while thofe people whofe anceftors were great, look with a fullen and inactive contempt on the flourifhing offspring of a later period. Europe has realized this idea, and it is now extended to America, where, from a few adventurers, a power is rifing, that it would be in vain to expect, could fpring from the more favourable climates of Greece or Italy; where it would appear that the people, as if fatisfied with the glory of their anceftors, are infenfible to the ftate of con-temptible indolence and infignificance into which they have themfelves fallen.

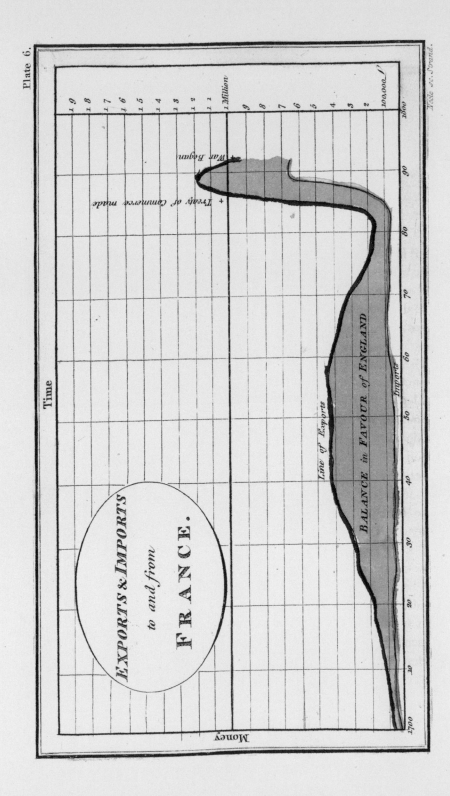

Plate 6.

EXPORTS & IMPORTS
to and from
FRANCE.

Time

Money

Treaty of Commerce made

War Begun

Line of Exports

BALANCE in FAVOUR of ENGLAND

Imports

1700 10 20 30 40 50 60 70 80 90 1800

100,000 l 2 3 4 5 6 7 8 9 1 Million 1 1 1 2 1 3 1 4 1 5 1 6 1 7 1 8 1 9

Neele sc. Strand.

OBSERVATIONS on the TRADE

with

FRANCE.

CHART VI.

WE have now before us a very fallacious reprefentation of the trade between two countries, which, from their fituation, as well as from the nature of their productions, we might expect to find immenfe; yet which, through a ftrange fpecies of policy, is extremely inconfiderable.

There cannot be a doubt that the illicit trade far exceeds in amount that here delineated, which can include only what is regularly entered. This trade furnifhes us with an aftonifhing inftance of the inefficacy of the laws that are injudicioufly enacted, and which furnifh too great a reward for evafion. When it is intended abfolutely to prevent the importation of an article altogether, then the higher the duties are the better; but, if *revenue* is the object, then the problem becomes much more difficult, and involves many different cafes, though in general it is found more deftructive of the end, to lay on duties that are too high, than fuch as are too low. Of the truth of this, our trade with France is a ftrong inftance; for the duties laid on by both nations, and the laws made, have counteracted and deftroyed themfelves. Nothing hurts the minds of men fo much as a temptation to do things in a concealed and hidden manner. It makes them at once acquainted with the modes of hurting mankind; they become eftranged from thofe

people

people whofe works bear the light of day, and they gradually proceed to every fort of immorality. Induftry is hurt
by whatever is unfriendly to virtue, than which nothing
is more fo than illicit trade; but it is alfo injured by the
idea of their being any other road to wealth, or even to
fubfiftence, than that of itfelf. Smuggling holds up to the
idle a method of getting a living, and perhaps of getting rich,
that is to them far more agreeable than that of regular labour.   Thus the duties laid on French goods are productive of other effects than thofe that are merely confined to the
trade itfelf; and therefore they merit the more particular
attention.

   The advantages that would refult to both countries, from a
liberal fyftem of commerce, are numerous; but there feem
to be many obftacles in the way.   That fame proximity
of fituation that would render the commerce fo very advantageous, has given rife to thofe laws that have nearly
deftroyed it altogether ; and the abfurd idea of letting
commercial concerns be regulated by that rankling animofity
which is fo confpicuoufly great between near neighbours when
they go to war, has at all times inclined either one or other of
the nations we fpeak of to oppofe any fyftem founded upon
a good underftanding, in the time of peace.   When two
individuals meet to make a tranfaction of any kind, they no
doubt muft always be fenfible that their interefts are oppofite;
but they do not on that account conclude, that either of them
is of neceffity to be injured by the tranfaction; nor does any
party neceffarily wifh it to be difadvantageous to the other:
but, when the two nations of which we now fpeak make a
commercial arrangement, they feem to be actuated by a very
ftrange combination of interefts and intentions; they wifh very
properly for their own intereft, and as improperly wifh to
                                                    prevent

prevent that of the other; and they forget that advantages that are not reciprocal are of fhort duration.

Upon the whole, there is not a thing more to be defired than a commerce with France, upon enlarged and liberal principles; nor would the advantages be lefs to that country than to this; we fhould both be greatly benefited by fuch an arrangement; added to the folid advantages that we may certainly expect, there would be fomething fo agreeable in doing bufinefs confidentially, with neighbours fo near to us, and whom we individually refpect, that it is peculiarly to be defired: and, as mankind underftand their intereft better and better daily, there probably will be a lafting treaty of commerce between England and France, when they come fully to underftand their own intereft; and when the prefent unhappy differences are at an end; for which moment every real friend to the welfare of either country muft fervently wifh.

The fhort period, during which the treaty of commerce was in force, has produced no proof either in favour of or againft a commercial connection: for,. in the firft place, that treaty was not on good principles; and, in the fecond, it exifted during a time in which the French government was going to ruin, and all private affairs were fuffering from the fituation of the public. On this fubject much is to be faid; but it would be unnecef- fary, and perhaps improper, to treat on it at prefent.

D OBSER-

Plate 7.

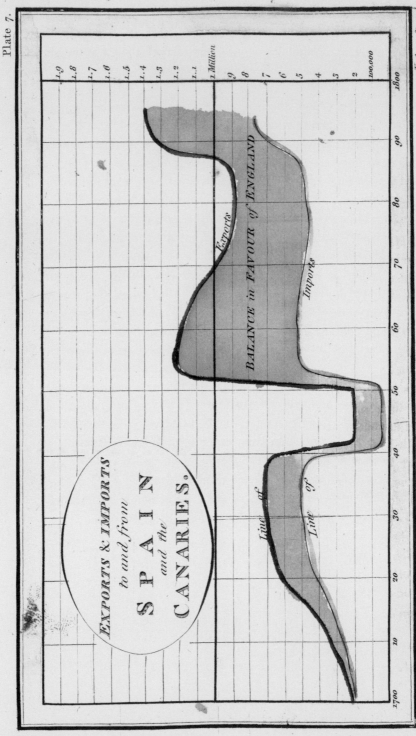

Neele sc. Strand.

OBSERVATIONS on the TRADE

WITH

SPAIN.

CHART VII.

THE trade with Spain, which is confiderably greater fince the laſt Spaniſh war than it had before been, is ſtill but ſmall, much more ſo than it probably would be, were it not for the unfavourable operation of treaties and alliances.

The Spaniards themſelves conſume and uſe the articles that they import from us; but they do not manufacture thoſe things which they export. They are indebted to South America, and to their good ſoil and climate for what they poſſeſs; for their own induſtry produces but little. Nature, as if it had intended Spaniards for idleneſs, has even furniſhed them with iron that forges without the trouble of heating in the fire; and they, on their part, are the moſt indolent ſet of men in the world.

The trade of Spain is not conſiderable to any country, nor do they pay much attention to it, except in gold and ſilver, which articles, though they may ſeem to deſerve a preference to other branches of commerce, do not merit it in reality; Spain has ſunk rapidly in the ſcale of nations, ſince the acquiſition of the riches of Mexico and Peru.

Among thoſe nations which, during the increaſe of the Roman empire, withſtood for a time its power, the inhabitants

of

of Spain made a moſt reſpectable figure as a ſteady, warlike, and hardy people.

The ſame period which brought peace to Spain, under the reign of Ferdinand and Iſabella, completed the deſtruction of commerce and manufactures in that country, becauſe it completed the expulſion of the Moors and the Jews; the moſt induſtrious, and at that time, when all arts were nearly loſt, the beſt artiſts in Europe.

The ſucceeding reign of Charles the Fifth, and the great inundation of the precious metals from Mexico and Peru, the treaſures of the Incas, all tended to aboliſh induſtry, and to confirm thoſe habits of pride and indolence for which the Spaniards have ever ſince been famous. Had the idea of a nation of gentlemen been capable of exiſtence, it would have been realized in Spain: their gold, their territory, and their diſpoſition, were all in favour of that extravagant notion; but unluckily, one gentleman requires a number of ſervants, and there the impoſſibility of execution puts an end to the idea; for the poor and low will always be the moſt numerous claſs in every country.*

The riches of South America, coming into Europe by the way of Spain, will, while they come ſo, occaſion a conſiderable trade of imports with other countries, to afford a market for the gold, which they exchange for the leſs precious, but not leſs uſeful manufactures and productions of other parts; and except when treaties ſtand in the way, no nation can ſupply them ſo well with what they want as England.

* It is impoſſible here not to make the remark, that the *ſame cauſe* prevented the Spaniards from being all gentlemen; that during the frenzy of the late revolution in France, prevented the ſyſtem of perfect equality, this *ſame cauſe* was the abſolute impracticability of both.

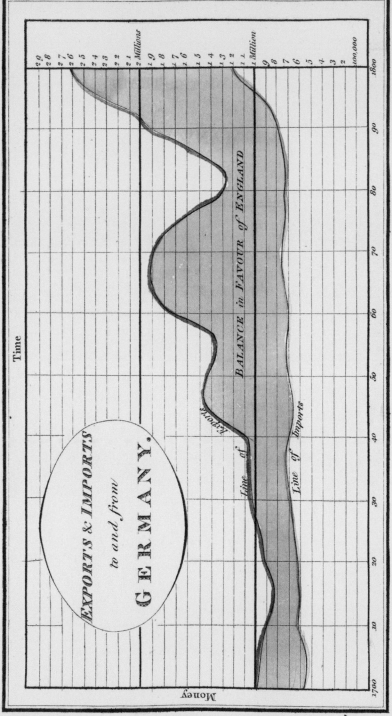

Plate 8.

Time

Money

EXPORTS & IMPORTS
to and from
GERMANY.

BALANCE in FAVOUR of ENGLAND

Line of Exports

Line of Imports

2.9
2.8
2.7
2.6
2.5
2.4
2.3
2.2
2.1
2 Millions
1.9
1.8
1.7
1.6
1.5
1.4
1.3
1.2
1.1
1 Million
9
8
7
6
5
4
3
2
100.000

1700   10   20   30   40   50   60   70   80   90   1800

Neele sc. Strand.

OBSERVATIONS ON THE TRADE
TO
GERMANY.
CHART VIII.

THE trade with Germany, very confiderable in its amount, is alfo from its nature one of the moft advantageous branches of our commerce. The ftrict honour and integrity that fo early diftinguifhed the individuals of that nation accompanies ftill all their mercantile engagements. Befides this circumftance, in all cafes fo defirable, the articles which we import and export are, in their nature, very advantageous to us. Thofe that we import from thence are chiefly raw materials, and our exports confift principally of finifhed goods, the value of which is derived from the labour and art in making; fo that they afford a greater advantage, and are a fource of more riches to us than twice the trade might be, if the articles were of a higher intrinfic value.

The articles exported to Germany are chiefly of the fort that the Germans manufacture themfelves. That country, which has frequently given both government and arts to modern Europe, and which to this day produces the very beft artifts, was unlucky in having ftrict laws made, relating to freedoms and corporations, at an early period, when the miftaken notion prevailed, that manufactures were improved and encouraged by fuch privileges and reftrictions. It has been owing to this circumftance that we have often fupplied them

with

with articles, the art and manner of making which had originated in their own country.

The Germans excel in goodneſs of work, but by no means in diſpatch of buſineſs. Individuals there are taught, from their firſt going to learn a trade, to conſider excellence of workmanſhip as the thing of all others the moſt deſirable to be attained. This diſpoſition in individuals has prevented them from manufaćturing cheap many articles which they can make much better than any other nation in the world. The nature of buſineſs in Britain, which is often calculated more for diſpatch and low prices, than for the goodneſs of the article, enables us to underſell them; and, particularly for theſe fifty years laſt paſt, during which our dexterity and improvements in arts have advanced amazingly.

The Emperor, Joſeph I. who was always awake to the intereſts of his country, counteraćted, as far as poſſible, and always diſcouraged thoſe bad monopolizing laws, which the German diſpoſition, naturally averſe to change, was not willing to have repealed; and perhaps if the ſame monarch had not, by his unwiſe, or at leaſt unſucceſsful, projećts in politics and religion, involved the peace and happineſs of his country, the German manufaćtures might in time have met us in our own way; but, as it is, that country is now ſo deranged, that we muſt be contented to ſpeak of the paſt without preſuming to gueſs about the future.

OBSER-

Plate 9.

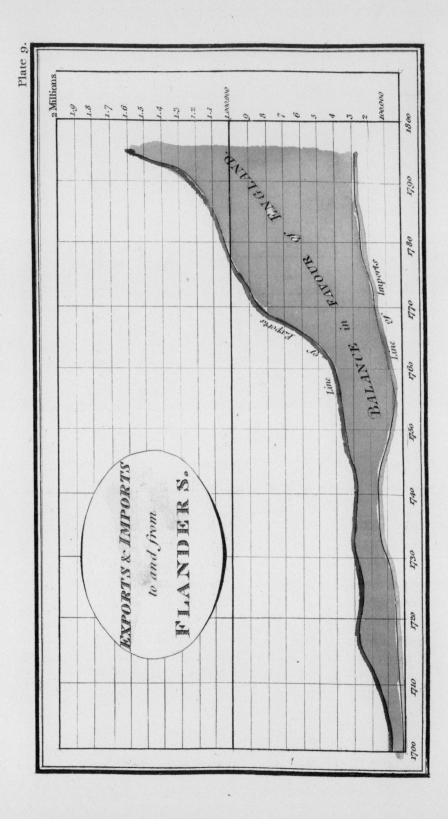

OBSERVATIONS on the TRADE

to

FLANDERS.

CHART IX.

THE trade to Flanders, which confifts chiefly of thofe of our manufactures that are neceffary in a polifhed country, has become a very confiderable, and a very beneficial one within thefe laft forty years.

The excellence of our hardwares, and many of our wove ftuffs, have been the occafion of this late increafe and balance in our favour. The imports from thence are very inconfiderable; for the excellence of our manufactures preclude the neceffity of our importing any of the great articles of expenfe, and they have not natural productions to furnifh us with to any confiderable amount, at leaft not fuch as we want.

Particular places will always excel in the art of fome manufactures, which are not worth while imitating in other parts, or which cannot, from the nature of things, be manufactured in every country, as there is not fufficient confumption to employ fo many different eftablifhments; except from this caufe, and fuch productions of the earth as are not to be found in England, we import fcarcely any thing but raw materials. There are other markets nearer at home, and better than England, for whatever Flanders has to fpare. Such a trade as this is of more real advantage than one of a much greater amount to

D 4

America,

America, which cannot pay for any thing foon, and often does not at all; and which opens a wild field for deception. Yet fuch is the difpofition of men, that we value what is fpeculative and precarious, more than what is fafe and beneficial. The fupport and protection of our trade to Flanders ought to be a matter of public attention, as it is of great public advantage.

Flanders is too rich an agricultural country ever to excel much in manufactures; for it is certain that manufactures do never fucceed fo well in a very fertile country, as in one where nature does lefs for the inhabitants.

OBSER-

Plate 10.

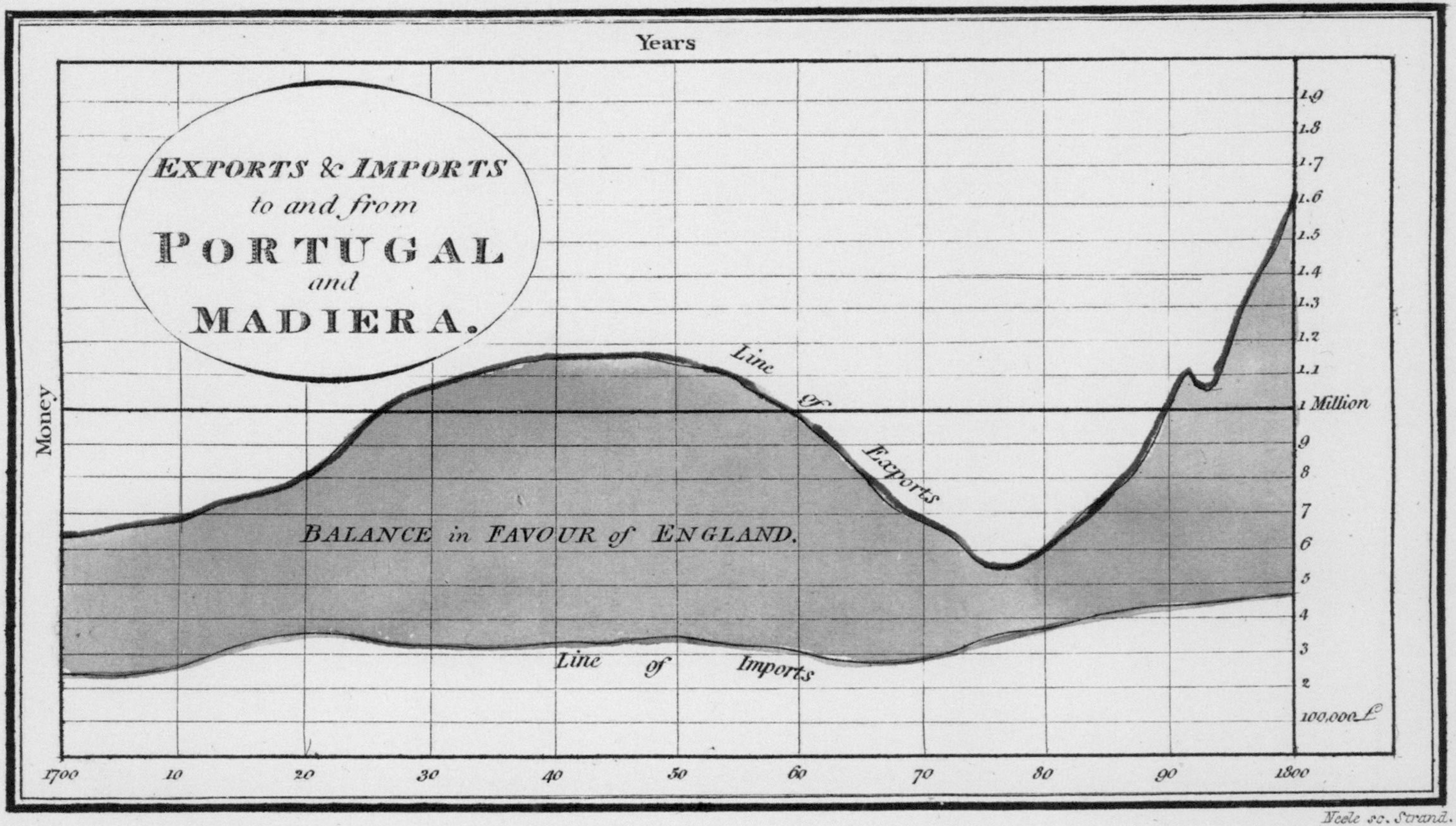

Years

Money

EXPORTS & IMPORTS
to and from
PORTUGAL
and
MADIERA.

BALANCE in FAVOUR of ENGLAND.

Line of Exports

Line of Imports

1.9
1.8
1.7
1.6
1.5
1.4
1.3
1.2
1.1
1 Million
9
8
7
6
5
4
3
2
100,000£

1700   10   20   30   40   50   60   70   80   90   1800

Neele sc. Strand.

OBSERVATIONS on the TRADE

to

PORTUGAL and MADEIRA.

CHART X.

THE difpofitions of the Portuguefe are not much unlike thofe of the inhabitants of Spain, with regard to commerce, only that they are worfe. We export a vaft deal more than we import, and our exports are the work of our hands, theirs are the productions of their country, or of their poffeffions; of which laft they have fome that are very valuable.

Thofe naval difcoveries, which have as it were altered the geography of countries, and the manners of men, were firft begun by the Portuguefe; but they foon gave up enterprize, and are now rather a paffive than an active ftate, both in commerce and in politics. They refemble thofe men, who, ignorant of the ufeful arts of life, fearch for fhells and pebbles, which, when found, they exchange for what are ultimately much more valuable, common neceffaries. It is not for their own ufe that they bring from all quarters of the world, what is moft valuable and moft rare; gold, pearls, and even diamonds, with the moft aromatic and grateful of perfumes and fpices, from Africa, America, and the Eaft. Thofe are referved for men who have learned to falt fifh, to raife corn, to manufacture the neceffaries and enjoy the luxuries of life.

The iflands of Madeira make part of the valuable poffeffions of the crown of Portugal, and furnifh fome of the moft

delicious

delicious wines in the world. The firſt ſugars that were in Europe are ſaid to have been produced in thoſe iſlands; but they are not now the cheapeſt, and have given place to the cultivation of the vine, in which thoſe iſlands may probably for ever remain unrivalled; for the particular flavour and variety of thoſe wines exempt them from that poſſibility of imitation, or that ſpecific price being fixed, that enables one place to ſupplant another in the market, for articles that are of leſs exquiſite nicety. Wherever there is much variety in quality there will be a latitude in price, either ariſing from taſte or faſhion, both of which have influence on the uſe of wines, and liquors of all ſorts. The caſe is very different with articles that poſſeſs not that variety; ſugar, for inſtance, bears a certain price, in proportion to its being finely manufactured; but people are not ſkilful in the taſte of it, and there is no faſhion in ſugar; ſo that the cheapeſt market is the beſt. This occaſions the ſame diſtinction among the produce of different countries, that is ſo well known to exiſt between mechanical labour and the fine arts.

Number, quantity, or quality, meaſures the value of the former with preciſion; but excellence, and the opinion of men, fix a more undecided value on the latter. The one ſort, therefore, very naturally ſettle where they can be done cheapeſt, and the other where they may excel moſt. Theſe two diſtinct characters, of the uſeful and the fine arts, intermix and divide in different degrees, and in a manner that gives riſe to infinite variety. Much entertainment, and great inſtruction, would ariſe from an inveſtigation of this, and of its conſequences, at length, as this diſtinction enters into the value of moſt things, and is intimately connected with the general principle of commercial affairs, but that inveſtigation would here be miſplaced and improper.

OBSER-

Plate 2.

Time

EXPORTS & IMPORTS to and from HOLLAND.

2.9
2.8
2.7
2.6
2.5
2.4
2.3
2.2
2.1
2 Millions
1.9
1.8
1.7
1.6
1.5
1.4
1.3
1.2
1.1
1 Million
.9
.8
7
6
5
4
3
2
100.000

Exports

BALANCE in FAVOUR of ENGLAND

Line of

Line of Imports

1800
90
80
70
60
50
40
30
20
10
1700

Money

Fecte & Strand.

OBSERVATIONS on the TRADE

WITH THE

SEVEN UNITED PROVINCES.

CHART XI.

THE republic of Holland poffeffed, at one time, the greateft part of the trade of Europe: its fifheries occupied the coafts, and its merchant-veffels the ocean. The Dutch were the venders of riches, and the carriers of goods to the reft of the world; and nothing that would bring a price, from a herring to a whale, was confidered as below their attention, or above their reach. With a moft aftonifhing degree of induftry and perfeverance, they raifed an inconfiderable fubordinate province to the rank of one of the moft powerful ftates in Europe; and they fhewed, that induftry is rather the child of neceffity than of opportunity. Upon the coaft of the Mediterranean fea, a number of fifhermen had raifed up, from a fea-weed bank, the elegant, the rich, and powerful republic of Venice; which for a time engroffed the carrying-trade of Europe; till another fet of men, upon another fwampy marfh, in a fituation lefs favourable indeed, but with an indefatigable obftinacy of difpofition, which, when impelled by neceffity, is almoft equal to any purpofe, wrefted from Venice the fuperiority in trade, and Amfterdam became the firft commercial city in the world.

The changeable nature of things, which often in a fhort time raifes ftates, as well as individuals, to the higheft pitch

of

of profperity, likewife finks them into a low and indigent fituation. Other nations have now opened their eyes to the advantages of commerce; and though they cannot all afpire to become carriers for others, yet they in general wifh to become carriers for themfelves. In proportion as this takes place, it is evident, the extent of Dutch trade and of Dutch confequence muft decline. The internal manufactures of Holland, though confiderable, have never been its chief fupport; and, as for the productions of a country fo limitted in extent, and well peopled, they never have fupplied itfelf; fo that it has been literally a ftore-houfe and nation of carriers, and confequently, its wealth cannot be very ftable, but muft exift by the indolence or want of fkill of other nations, and muft fall by their exertion and induftry.

The Dutch, although they want territory at home, have had the addrefs to fecure fome very valuable poffeffions abroad, and they did till very lately derive much advantage from the exclufive poffeffion of the Spice-Iflands in the Eaft Indies.

The trade between this country and Holland before the war was lefs than it had been, partly perhaps, becaufe the general amount of the trade of the ftates is diminifhing; and partly, becaufe we carry our own manufactures directly to the places where they are to be confumed. As the Dutch always buy at the beft market, and as they have the wifdom to overlook any difagreement where their intereft is concerned, we fhall at all times be fure of fupplying them with thofe articles which they can buy to better advantage here than elfewhere. This is pretty certain, for no laws, human or divine, nor the moft ftrict treaty, will make a Dutchman buy at a dear market, when he knows where to buy at a cheap one.

Whatever

Whatever it may be owing to, it is a certain fact, that the Dutch are the beſt fitted for mercantile buſineſs of any people in the world. Their paſſions are all ſubſervient to the love of gain. Enabled by this, and impelled by neceſſity, they have certainly done wonders; for, beſides that the land is inſufficient in extent to ſupply the inhabitants, the expenſe of fortifying it from the ſea, as well as that of reſcuing it from the dominion of Spain, have loaded them with heavier taxes than any other nation. The powers of getting money were fully called forth by nature and circumſtances in Holland, and they have exerted them to the very utmoſt hitherto; though undoubtedly, that energy by which they were once enabled to ſucceed ſo well, muſt leave them. Their unanimity is already gone, and their frugality diminiſhes: ſo that in time they probably will be reduced to that rank among nations, to which they are only entitled by numbers, by territory, and by wealth. It is unneceſſary to endeavour to enumerate the articles of Dutch imports and exports, for they deal in every article that is known as a branch of trade.

The monied capital of the Dutch is ſuppoſed to have ſuffered a great diminution ſince the French invaded their country; their ſituation has, in many other reſpects, become leſs favourable than heretofore for credit and exertion of induſtry. How all this will operate on our future commerce with that people time only can ſhew; for the events of the preſent period are unlike to every thing of which hiſtory has furniſhed any example.

Plate 12

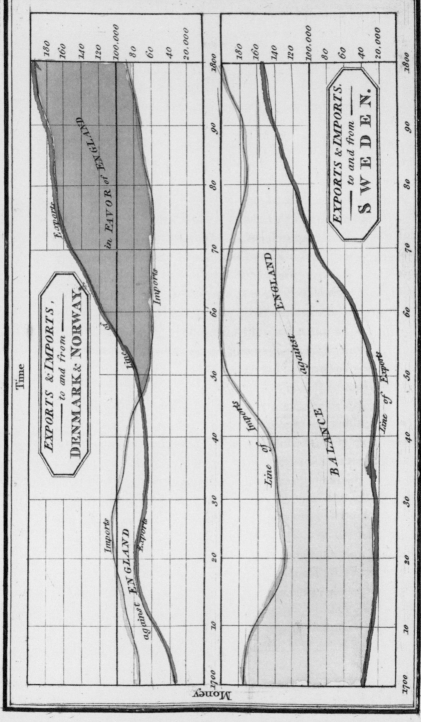

OBSERVATIONS on the TRADE

WITH

SWEDEN.

CHART XII.

THE articles of our imports from Sweden are numerous and valuable. Of these, iron, hemp, flax, and pitch, are the most considerable. The iron of Sweden is sold in a state more generally marketable than any other iron, from its shape, and the variety of sizes of the bars; and from its excellent quality, it also is of more general utility. Upon the subject of the iron and steel trade in general, the best information extant, in this, or perhaps in any country, is to be found in Lord Sheffield's Observations on Trade and Commerce.

His lordship distinguishes, with great propriety, between the duty on imported iron, and duties on other raw materials, which in general are very hurtful to manufactures; but on account of this being a manufacture that promises to succeed in this country in a high degree, that duty on the foreign article serves as a premium on the home manufacture.

Of the iron imported in quantities in this country, the assorted Swedish is by far the best, as well as most convenient for use; and people in the iron trade should attend to this circumstance. A considerable part of the iron (excepting for shipping and foreign manufactures,) is used by men who are in a very small way of business in country towns, who, having no money

12

to lay out on a large ftock, buy the kind that will do for any chance purpofe. Now the Swedifh iron is exactly this fort; for with a few bars of that, well chofen, a poor country black-fmith will be better fupplied than with five times the quantity from any other market. It is to be obferved, that they are all uncertain, when they purchafe it, what particular ufe it is to ferve, and therefore they muft have the kind that will do for any of the probable purpofes. This is well worth the attention of iron-mafters, for it is furprifing how much demand always has been occafioned for Swedifh iron, by that fingle circumftance of the inequality of the bars.

It is likely that home-made iron will fupply the place of Ruffia iron firft, as that is not equal in quality to Swedifh ; for, from the nature of any new art, it is eafier to rival the leaft perfect firft ; and a reduction of price is more eagerly grafped at by the manufacturers who ufe the coarfe, than thofe who ufe the fine material, becaufe the quantity and value of it generally bears a greater proportion to the whole value of the goods. Thus, for inftance, in wire, or in fine wood fcrews, the material does not in many cafes make one-tenth of the value ; but in large bolts, and large nails, it makes fix-tenths : fo that a faving of five per cent. in material of the latter, is of fix times the advantage that it is in the former, fairly and accurately : but in reality it operates more powerfully than in the proportion of fix to one, for the fix will always be confidered as an object defireable to be faved, though the other may not be confidered as worth thinking about at all ; it is a miftake, to think that things are always regarded and valued according to mathematical or numerica quantity, as it would appear to a reafoning theorift they ought to be ; for, to the mind, as well as to the eye, there are quantities too fmall to be perceived or noticed.

The

The different qualities of iron, the cold-fhort, the red-fhort, &c. &c. &c. feem to be produced, not by the original material, but by the method of manufacturing. And if fo, a few years will produce iron of every kind, and of the *very beft* of every kind in England; for the number of ingenious and liberal fpirited men, who employ their talents and their capitals to the improvement of that manufacture, was perhaps never equalled in any age, or in any country. But it is an art that does not, like a mechanical contrivance, admit of being brought to a very perfect ftate at once; becaufe obfervation and experience, both of the men and their mafters, are the guide, and not reafon or invention. With regard to the exports of Sweden, they are few and inconfiderable; luxury has not made a great progrefs in that country; when it does, Britifh mannfactures will be its attendants, as they generally are to other parts of the world.

OBSER

OBSERVATIONS on the TRADE

to

DENMARK and NORWAY.

CHART XII. NO. 2.

THIS trade has changed greatly within thefe fifty years. Our imports are much the fame in amount with what they ufed to be, but our exports are much higher. This may be owing to the nature of the articles, what we import being materials, and our exports manufactured goods; for it is generally to be obferved, that a trade that confifts of raw materials, or unfinifhed produce, is more regularly of the fame amount than one that confifts of articles in a finifhed ftate; particularly if thefe laft are articles of luxury and expenfe. That is the cafe with this branch of our trade; what we export there being chiefly manufactured goods, and our imports from thence confifting of raw materials; for we import little from thence that is in a confumable ftate. The amount of trade altogether is fo inconfiderable, that without any great change in the ftate of either country, it might rife or fall one half in its amount; nor can it be any caufe of wonder, that Denmark fhould import thrice the value that it did half a century ago, when we confider how much our manufactures have improved during that period.*

---

* The revenues of Denmark have confiderably more than doubled fince the year 1630; fo the country probably is getting richer very rapidly.

Manu-

Manufactures flourish inconsiderably in Denmark. The rude art of producing in a saleable state, tallow, hides, pitch, and tar, with the more difficult but well known art of making iron, sums up nearly the ingenuity of that country; and these articles, with timber, chiefly compose the exports to other parts.

During the present war, Denmark has, as a neutral nation, enjoyed great advantages, all of which it has just now abandoned, either from fear or ambition, but how far indemnity or advantage can be obtained even by the fullest accomplishment of the claims concerning neutral vessels, it is very difficult to determine.

Plate 13

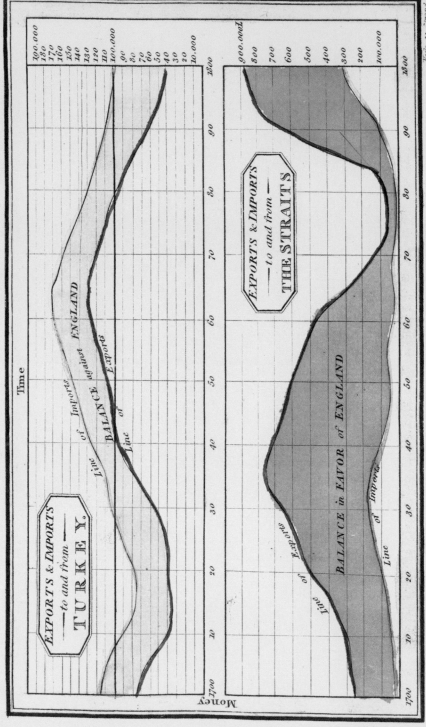

Time

EXPORTS & IMPORTS
— to and from —
TURKEY.

Line of Imports against ENGLAND

BALANCE

Line of Exports

700.000
180.
170.
160.
150.
140.
130.
120.
110.
100.000
90.
80.
70.
60.
50.
40.
30.
20.
10.000

1800
90
80
70
60
50
40
30
20
10
1700

Money

EXPORTS & IMPORTS
— to and from —
THE STRAITS

BALANCE in FAVOR of ENGLAND

Line of Exports

Line of Imports

900.000
800
700
600
500
400
300
200
100.000

1800
90
80
70
60
50
40
30
20
10
1700

Neele sc. Strand.

OBSERVATIONS on the TRADE

to

TURKEY.

CHART XIII.

ONE of the fineſt and the faireſt portions of Europe, where arts, ſcience, and literature, once flouriſhed in a high degree, is now poſſeſſed by the moſt ignorant, indolent, and debaſed race of men that ever incumbered the face of the globe: and the ſame ſpot which was famous for giving birth to the firſt of orators, poets, and philoſophers, is now filled with a wretched and contemptible ſet of mortals, who groan under one of the moſt miſerable, deſpotic, yet feeble governments that ever had exiſtence.

The ſituation of Conſtantinople, the capital of Turkey, choſen by a Roman emperor as the fineſt in the empire, is alſo one of the fineſt in the world. Its ſituation for trade was unexceptionable, and its buildings were the moſt magnificent; the ſurrounding country abounded in every thing that the moſt fertile ſoil or grateful climate could produce. Yet the nature of the government, and inhabitants of the country, which ſometimes produce a garden upon a rock, have here reverſed the caſe, and converted a moſt noble country into one, poor and deſpicable.

When the hand of gothic and of ſavage barbarity extinguiſhed throughout Europe the flame of ancient learning and

E 3   the

the arts, a fingle fpark was left unextinguifhed in Conftanti-
nople, which had not quite perifhed, when once more the
dawn of light arofe; fo that this place became a fort of feeble
link between the ancient and the prefent world: and fome
arts were preferved, which are peculiar to that feat of ignorance, even to this day; particularly thofe of dying fome co-
lours, which were preferved there; and, being of a nature
eafily to be kept a fecret, have never either been with fuccefs
difcovered or imitated.

Pride and meannefs are blended in the characters of the
worfhippers of Mahomet, in a peculiar degree.   Though they
fubmit to be flaves, yet they think it below their dignity to
follow commerce in the ufual way; and other European nations are obliged to go to them for thofe articles, which an induftrious and mercantile nation would fend out at their own
rifk, and in their own fhips: fo that they underftand little of
naval affairs, and their commerce is much lefs than it would
otherwife be.   To what particular circumftances it is owing,
that their commerce declines, may not be eafily accounted
for, with any degree of certainty but it is probable that it
may be with juftice attributed to our rivalling them in thofe
arts in which they originally excelled fo much, and ftill do
excel other nations; for we ufe more fine carpets and Turkey
leather than at any former time; but many are imitations of
our own manufacture.   Some fine cotton ftuffs, of an excellent dye, are alfo imported from thence; fome of the produce
of the ground, but not many of their manufactures.

The Turkifh empire feems now on the eve of undergoing
fome changes that will probably alter the habits and manners
of the people, after which the probability is, that the commerce with that country will be greater than ever.

OBSER-

OBSERVATIONS on the TRADE

TO THE

STRAIGHTS.

CHART XIII. NO. 2.

THE remarks that this branch affords are very few; unlefs we were to deviate from the plan that has been followed with regard to the others.

The fmaller and more inconfiderable branches of our trade will ever be more fluctuating than the large and extenfive ones; and they feem to have frequently, during this century, gone by a different rule; for while the greater branches, with very few exceptions, have increafed, thofe have frequently been upon the decline. Of the truth of the latter part of this affertion, the chart now before us is a proof.

The coafts of the Mediterranean fea were originally the feats of commerce; and for many obvious reafons, until navigation began upon the great and extenfive fcale, when men croffed the Atlantic, and doubled the Cape of Good Hope, they were the moft favourable for commerce in the world. The rich and powerful ftates were all upon thefe coafts; and it was the way by which the luxuries of Afia came into Europe. Though this ceafed to be the cafe, and that though that time was at an end, before the commencement of the period we are now confidering, yet there were many veftiges of the ancient wealth, that are gradually wearing away. The

E 4 civilized

civilized part of mankind was upon a very fmall fcale, from the fall of the Roman empire till the fifteenth century; and while that continued to be the cafe the places which were beft fituated, got the greateft fhare. Religion, and the love of adventure, both operating at once, in the eleventh century, upon men's minds, drew moft nations of Europe to the Holy Land, in order to extirpate the infidels: one confequence of which was, that whatever moveable wealth was poffeffed by thefe barbarians, was carried to that part of the world, which grew rich at the expenfe of the enthufiaftic multitude. Venice rofe into a rich republic by that wealth, and leffer ports fhared, in a fmaller degree, the fame advantage.

This wealth, added to the fine fituation, completed the advantage that the ports on the Mediterranean had before enjoyed; and as time is neceffary to make countries poor, as well as to enrich them, the trade only declined by degrees; and in proportion to the commerce of the reft of Europe, it ftill continues to decline. The trade to Italy and Venice, which has been treated of feparately, and is increafing, may perhaps occafion an unfair conclufion alfo; for, lying fo conveniently as it does for an intermediate market, the other bufinefs up the Straights may frequently be tranfacted there; as all mercantile people and ftates purchafe, as well as vend, things that they themfelves neither produce nor confume.

OBSER-

Plate 14

EXPORTS & IMPORTS
to and from the
SPANISH WEST INDIES.

EXPORTS & IMPORTS
to and, from the
B A L T I C.

OBSERVATIONS on the TRADE

to the

EAST COUNTRY and BALTIC.

CHART XIV.

THE balance of this trade is greatly againſt us; but we only purchaſe from thence uſeful materials, that we again work up into much greater value. So that although the balance is againſt us, we are not any loſers by it. There is a great deception in looking merely at the balance in money, which is certainly not the true and only meaſure of the advantage, or loſs, with which trade is attended. The nature of the goods enters very much into the caſe; and when they are raw materials that we purchaſe, it is frequently as good as if we were paid in caſh. It is always ſo, when we are paid in ſuch raw materials as we cannot produce to much advantage at home, and this is a good deal the caſe with the articles we import from theſe northern countries. Though we raiſe flax at home, and though in the end there may perhaps be a great advantage attending it; yet there is not, at preſent, very great advantage attending it; nor are there any things that we do purchaſe there, that are not at ſo reaſonable a price in time of peace, that we could ſcarcely produce the like at home ſo cheap. The commerce to that country is therefore a good one, though we ſeem to loſe by it; and there is this difference between a trade where we pay a balance, inſtead of receiving one, that we are more ſecure of its continuance.

When

When we begin to improve either manufactures or lands, in this country, we account it certain, that, if we raise or make the articles as cheap as we have hitherto bought them, the business is sure to succeed.   But it frequently happens, that the people who supplied us before are able to reduce the article in price, rather than be undersold; and it is not unlikely, that this is the case with those parts.   In that event, we certainly gain a point of importance, reducing the price; but it is not the point for which we wished.   This is more likely, by much, to be the case with the productions of land than with manufactures; for the produce of ground (particularly in fine rich countries) admits of a greater reduction; as the rent of land would in that case be lowered, till its productions became saleable.   There are a great many attending circumstances upon which this depends.   In case corn and grass can be raised on these lands to advantage, or if sale can be got elsewhere for flax, it will not be reduced upon our raising flax at home.   If, on the contrary, the grounds will not turn to advantage into corn or grass, then shall we have the price of flax and hemp reduced, till it is our interest to become purchasers, rather than to raise it at home, where we can, with advantage, employ our ground to other purposes.

OBSERVATIONS on the TRADE

WITH THE

SPANISH WEST INDIES.

CHART XIV. NO. 2.

THIS feems the moft variable of any branch of trade that has yet come under our confideration; fometimes confiderable, and fometimes invifible.

Were there not fome mutual advantage, and a very confiderable one, the amount would not have been fo great fome years; neither, unlefs other circumftances took place befides the direct trading intereft, would it fometimes entirely ceafe. The fact certainly is, that there is no great nation, nor inconfiderable body of men, who do not wifh to trade with England. Thofe who cannot pay ready money, can get that length of credit here which they could get no where elfe; and thofe who have money, can lay it out to more advantage here than in any other place. A cheap market is always the object of the latter, and long credit of the former; and mankind is compofed, almoft entirely, of thefe two claffes; and therefore England is, for certain articles, either directly or indirectly, the manufacturer and ftore-houfe of all nations. And it probably is owing to this caufe, that in favourable moments, the trade with the Spanifh Weft Indies rofe, fuddenly difappeared, and rofe again, according to opportunity, which will be the natural cafe with any branch of trade that is beneficial in itfelf, but forbidden by the laws and commercial regulations. It is

advantageous

advantageous, in many different refpects, for a nation to monopolize the carrying-trade of her own colonies, and even the indolent Spaniards are ambitious of that advantage: but their flow habits, and little turn for commerce, prevent them from doing it effectually.

OBSER-

EXPORTS & IMPORTS
to and from
RUSSIA.

Line of Imports

BALANCE AGAINST ENGLAND

Line of Exports

1 Million

100,000 £

1700   10   20   30   40   50   60   70   80   90   1800

OBSERVATIONS on the TRADE

WITH

RUSSIA.

CHART XV.

THE Ruffian empire, which was fcarcely known to the other powers of Europe till the middle of the fifteenth century,* has not hitherto had wealth enough to fpare to be able to purchafe great quantities of goods from any country. The extent of the empire of all the Ruffias will have a curious effect upon its trade, for it will operate in two ways. The circumftance of its extending fo far from north to fouth, and by that means having all forts of foil, and varieties of climate, within itfelf, and producing almoft all the things that are neceffary to a nation, muft render its neceffity of importing to be lefs than that of a fmaller territory.

Again, that extent of territory will act unfavourably in many particulars. The government, for one thing, muft be adapted to the extent, and therefore muft be arbitrary, or, if not, it will be very ill obeyed; either of which prevents the fecu-

---

* In the reign of Queen Mary, an embaffy was fent from the Czar. The ambaffadors were wrecked on the coaft of Scotland, where they were hofpitably entertained, and proceeding to London were well received, (Hollingfhed, page 732.) This feems to have been the firft intercourfe which that empire had with the weftern potentates of Europe, (Hume's Hift. vol. iv. page 447.)

12   rity

rity of individual property, and will therefore never allow
arts, manufactures, or commerce, to flourish in any high degree. The commodities, therefore, that Ruffia will export,
will probably always be the natural productions of the country;
and its imports will chiefly confift of thofe articles, the manufacturing of which fucceed beft in well regulated countries,
like England. In proportion to the fize of the country, however, the trade will always be inconfiderable; for, if it were
neceffary, Ruffia is fufficiently various in productions to do
without importing any thing, almoft; and it never will be
(taking the whole nation together) a very polifhed and luxurious people. The court, indeed, and a fmall portion of the
empire, may be more magnificent and more luxurious than
any in Europe; for it has a greater extent of territory from
which to be fupported; but its extremities will never flourifh
highly; nor will all the vigour of the defcendants of Peter the
Great be able to fpread wealth and induftry, with their happy
effects, through the diftant extremities of fo extenfive a portion
of the world.

This trade, although it is long fince a Ruffia company was
eftablifhed, has, till within thefe fifty years, been but very inconfiderable. Our imports from thence have increafed rapidly, but our exports very little; perhaps for this reafon,
that, even at the beginning of that period, the court at Muf-
cow was magnificent and luxurious, nearly as much fo as it is
yet, and therefore probably wanted many of our manufactures;
but the extent of country wants none even to this day; for, if
it did, what we fend over is fcarcely fufficient to furnifh every
peafant in that extenfive empire with a knife to cut his meat.
But, although the confumption of articles of luxury has not
extended much beyond the ufual limit of the great cities and
the court, yet the production of goods for exportation muft
have

have increafed very confiderably under the aufpices of a fuc-
ceffion of able fovereigns, who had the good of the country
much at heart.   It is probable, from this, that the balance of
the trade of Ruffia will be in its favour with almoft every
other country, as well as with England; and, therefore, that
they muft be getting richer.   The natural confequence, in a
fmall country, would be, to become luxurious ; and perhaps
even *they* may do fo, at the end of a long period ; but it will
be a long one; for, at this time, the improvements neceffary
that they may avail themfelves of the advantages of fome parts
of their dominions, and that they may counteract the difadvan-
tages of others, will fwallow up whatever balance may be in
their favour, for many years, even for centuries yet to come.
When that progrefs comes to a period, then may we perhaps
have a balance that is not fo much againft us.

The Ruffian empire bids fair to fhine, in fome future
period, as a warlike people ; but, for its commerce, there is
not much to be faid.   Though the fhort time that has elapfed
fince Ruffia began to make a figure in the world, does not
furnifh us with a full proof of this, yet the prefumption is in
its favour ; for, even in early times, its commerce * has been
ftill

* In the year 1569, the Czar John Bafilides, who was a great tyrant,
gave to Queen Elizabeth an exclufive patent to the whole trade of Muf-
covy, (Cambden, page 403), and fhe, in return, agreed that, in cafe of
a revolt of his fubjects, he fhould have a fafe retreat in England.   After
the death of John Bafilides, his fon Theodore revoked the patent.   During
the exiftence of that patent, the Englifh carried goods along the river Dwina,
in boats made of a fingle tree, which they towed up the ftream as far as
Walogda.   From thence they carried goods feven days journey over land
to Yeraflaw, then down the Volga to Aftracan, where they built fhips,
croffed the Cifpian fea, and fold their manufactures in Perfia.   This was

3   a very

ftill more infignificant than its confequence as a nation: and as a farther confirmation of this opinion, we may juft confi- der what *all the Ruffias united* would be, were they as well cultivated and civilized as England, and if manufactures flourifhed equally under a free government. The immenfe empire would, in that event, be able to fwallow up all the other powers of Europe, in cafe of war; and, in times of peace, might have every neceffary, and almoft every article of luxury, without importing a fingle cafk of goods from any European nation.

a very bold mercantile adventure; but, from the difficulty and difcourage- ment, was never renewed, (Cambden, page 418). This happened about fifteen years before we had any trade with Turkey, and the eftablifhment of that company.

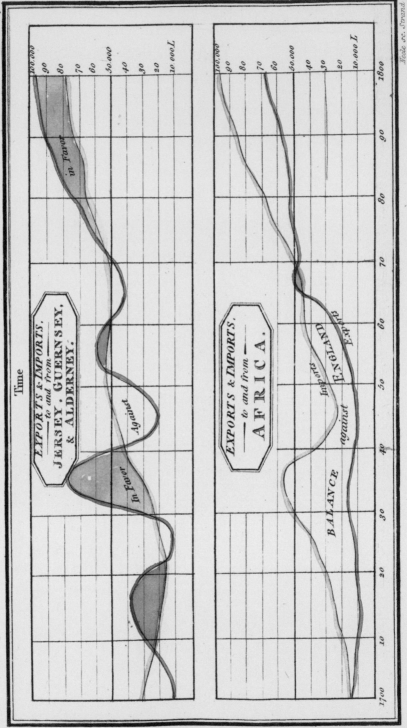

Time

EXPORTS & IMPORTS.
to and from
JERSEY, GUERNSEY,
& ALDERNEY.

in Favor

Against

In Favor

100.000
90
80
70
60
50.000
40
30
20
10.000 L

1800
90
80
70
60
50
40
30
20
10
1700

EXPORTS & IMPORTS.
to and from
AFRICA.

Imports

ENGLAND

Exports

against

BALANCE

100.000
90
80
70
60
50.000
40
30
20
10.000 L

Neele sc. Strand.

OBSERVATIONS ON THE TRADE

, TO

AFRICA.

CHART XVI.

THE nature of this trade, certainly not the moſt honourable in the world, affords room for much inveſtigation and remark in a moral or humane point of view : in a political or a commercial light it is perhaps leſs conſpicuouſly an object of attention. It conſiſts chiefly of commodities that are conſidered as holding a firſt rate place in the animal and the mineral world, for which in return the Africans receive the moſt raſcally articles that the ingenuity of Europeans has found means to produce. In return for our fellow creatures, for gold, and for ivory, we exchange the baſeſt of thoſe articles that are ſuited to the taſte or the fancy of a deſpicable ſet of barbarians. Whether the ſpirituous liquors or the fire-arms that are ſent there are moſt calculated for the deſtruction of the purchaſers, might become a queſtion not very eaſy to determine. The noxious quality of the one is at leaſt equalled by the danger attending the uſe of the other. There does not ſeem to be that regard to honour in this trade, which ought to make part of the nice character of the Engliſh merchant, unimpeachable, and unimpeached, upon the 'Change of London or of Amſterdam. It ſeems as if we kept our honour for ourſelves, and that with thoſe barbarians (who are more our inferiors in addreſs and cunning, than perhaps in any thing

F   elſe)

elfe) no honour, humanity, or equity, were at all neceffary.

The nations of Europe who have diftinguifhed themfelves
by the appellation of civilized, and of Chriftians, have uniformly fhewn this bafe difpofition; and there is not more
reafon for faying that hawks will kill pidgeons, than that
Europeans make free with the lives, property, and poffeffions
of the natives, whether of India, Africa, or America. Nobody will fay that if the hawk could eat the pidgeon, without
putting it to death, " it would not do fo." And if an European can enrich himfelf without injuring the natives of thefe
defencelefs countries, he is willing to do it; but we are un-
acquainted yet with any crime or any cruelty, that would
over-balance the love of gain in the breaft of Europeans,
when dealing with defencelefs natives, whether they are the
more civilized inhabitants of Peru, the irafcible ones of Africa,
or the mild natives of the Afiatic world: thofe who reap the
fruits of their rapine, meet with the honour and regard that
gold is generally fure to purchafe ; and the induftrious at home,
who acquire wealth by a more honourable means, fhare in the
ignominy thrown upon their country, without having themfelves fufficient regard to national charaĉter, or to juftice, to
put the mark of difgrace and averfion upon fuch aĉtions.

The advantage that arifes from the cultivation of our Weft
Indian plantations, by means of flaves, is much difputed. It
has been affirmed by many, that the labour of flaves is more
expenfive than that of hired fervants ; and there is reafon to
believe that it is fo ; for, from the nature and conftitution of
men, we may be well convinced that no punifhment can operate fo effeĉtually as the love of gain. Men are often found
in that ftate when they are hardened againft the fear of pain,
but

but never where they are indifferent to the pleasures pur-
chased by money.   Another thing is, that the most arbitrary
master cannot give a willing mind ; and who is there among
men who does not know, that the degree of labour of which
the human body is capable, is regulated full as much by the
temper of the mind, as by the strength of the body ?

In this country, houses for labour, or the works to which
notorious criminals are confined, never pay the expense, because
they are brought in competition with the *labour of free men.*
Were the labour of slaves, in like manner, opposed to that of
free men (seasoned and suited to the climate) it is probable it
also would be found not to pay.   Natives of Britain probably
cannot undergo that labour which is necessary to cultivate
plantations in hot climates; besides, they do not leave this
country with an intention to labour at the meaner employ-
ments.   But our islands are different from other parts of the
world, if they require a continual supply of new people from
other countries.   Allowing, however, that they do require a
supply, it cannot be so considerable as to oblige great numbers to be imported every year, if they were properly treated.
It may perhaps be said, that good treatment will not do with
negroes; at first it might not, but they would soon be convinced that it was their interest to behave well, and they
would do it.   No man would wish to emancipate the whole
slaves in an island at once, but it might be done by degrees,
and laws might previously be made for the purpose of regulating them, when made free.   Perhaps it might be necessary
to prevent them from ever purchasing land, and to lay them
under some other restrictions of that kind: perhaps also it
might be necessary to make a different set of punishments for
crimes committed by them or their descendants; and if it
were expected that money or territory were to be the reward

of this experiment, it would foon be made, in this age of adventure; but as that might not be the cafe, and as humanity
is the great argument, it is likely that it will not have much
weight with thofe people, on whom alone its operation is of
any importance.

OBSERVATIONS on the TRADE

TO

GUERNSEY, JERSEY, and ALDERNEY.

CHART XVI. NO. 2.

THIS branch of our trade, which is but fmall in amount, when compared with moft of thofe which have been reprefented in this work, is not fmall when we confider the extent of country, and the numbers of inhabitants in thefe little iflands, who are very induftrious.

It muft occur to every perfon naturally, that the fmaller a diftrict or an ifland is, the exports and imports will be the greater when compared with the number of inhabitants. Take the exports and imports of all Europe with the other quarters of the world, confidering Europe as one country, and it will not be found to amount to one fhilling a perfon per annum. Take the amount in Britain, it will be found about forty fhillings a perfon. Confider what is bought and fold by a fingle village, and it will be ftill greater than that. And laft of all, a fingle labouring family buys all that it ufes, and fells all that it produces; and the meaneft family taken in this way, does, proportionally, more buying and felling than the richeft ftate taken in a body. Confider the whole earth as one ftate, and it neither exports nor imports.

The internal trade, and external, which include all the foreign and domeftic tranfactions of men, amount together, to

the

the whole wealth produced by any nation, diſtrict, or family. As theſe two quantities compoſe a third, the increaſe of one of the quantities muſt be attended with the diminution of the other. Thus, ſuppoſe there is as much uncultivated ground in England, as, if cultivated, would grow as much flax as we import, then ſhould our imports become leſs by all the flax, and our internal trade would increaſe; ſo that to form a real eſtimate of the wealth of a country, a vaſt number of things come in. The exports and imports form a good baſis for the inquiry, but muſt be compared with thoſe other circumſtances of internal commerce and population. In treating of Ruſſia, it was obſerved, that it might be opulent and luxurious without any foreign trade at all, on account of its great extent, and variety of ſoil and climate; and farther, that at preſent its commerce is inconſiderable, when compared with its extent. Every perſon who pays attention to theſe ſubjects, muſt ſoon know this fact, and it is of particular conſequence that it ſhould be always kept in view, elſe they will be liable to make very falſe concluſions.

When ſome old facts are told, without conſidering all theſe things, England ſeems formerly to have been a very mean, poor country;* but when they enter into the calculation, we find that part of its great trade is occaſioned by actual wealth, and part by different habits of life; for ſuppoſe we ſhould chooſe to live without the productions of the Weſt, as we did four hundred years ago, and prefer Engliſh cloth and Engliſh

* The chief juſticiary of England, Jeoffery Fitzpierce, gave the king two *good Norway hawks* for a licence for Walter la Madine, to export one hundred pound weight of cheeſe out of the country. See Hume, Appendix II. page 133. And this was only a common tranſaction, and the manner in which the revenues of cuſtoms was originally inſtituted and levied.

food

food to the effeminate and luxurious fabrics and productions of the Eaft; our own malt liquors to the wines of France and Spain; our trade would be back at a very low ebb in a fhort time; though it is poffible that even then we might be internally very rich, powerful, and luxurious in our own way.

To reafoning like this, which applies to all commerce whatever, more or lefs, we muft attribute the amount of this comparatively confiderable trade to thefe fmall iflands.

OBSER-

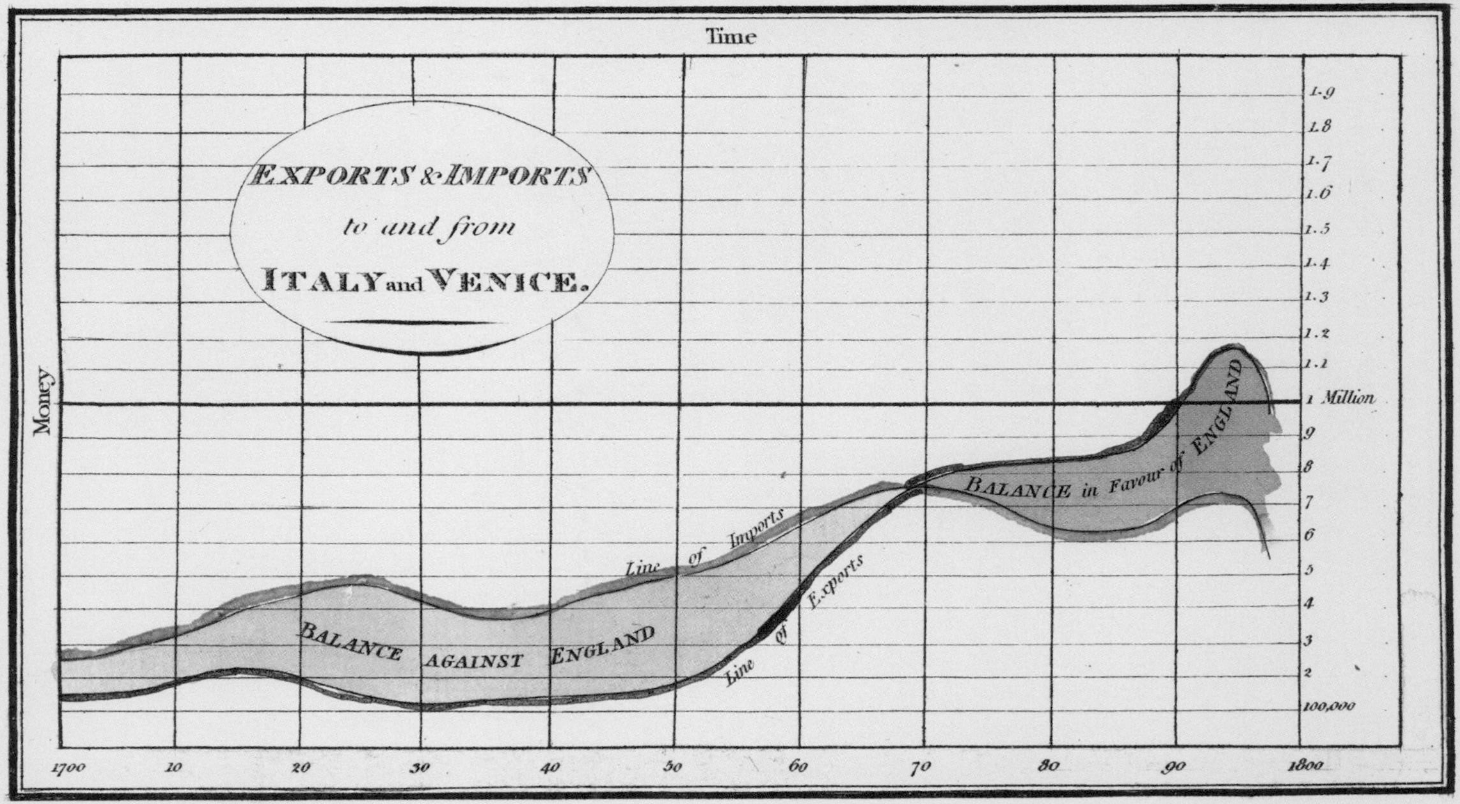

Plate 17

Time

Money

EXPORTS & IMPORTS
to and from
ITALY and VENICE.

BALANCE in Favour of ENGLAND

BALANCE AGAINST ENGLAND

Line of Exports

Line of Imports

1 Million

1700 10 20 30 40 50 60 70 80 90 1800

100,000 2 3 4 5 6 7 8 9 1.1 1.2 1.3 1.4 1.5 1.6 1.7 1.8 1.9

OBSERVATIONS on the TRADE

WITH

ITALY, including VENICE,

CHART XVII.

THERE is scarcely any branch of trade that admits of fewer remarks than that carried on between this country and Italy.

The commerce, which is not great in its amount, is not increased by any artificial means. The naturally fine, luxurious, and elegant productions of that country, will always oblige every polished and luxurious nation to apply there, either directly or indirectly, for those things which no other part of the world can furnish in an equally perfect state; and the indolent habits of men who boast of being descended from the masters of the world, will lay them under an equal necessity of applying to the less noble but more industrious part of mankind.

No country in the world has undergone so many reverses as Italy. After first emerging from insignificance, by subduing all its neighbours, and after being ruined by the luxury occasioned by their spoils, and falling a sacrifice to its own wealth and glory, it exalted itself a second time, and governed by religion the minds of men, with a more despotic sway, than it had by arms governed their persons. This second reign having also come to an end, the inhabitants of that country,

famous

famous for arms, religion, and the fine arts, cannot ftoop to the drudgery of common induftry, and it is now inferior to moft other nations in thofe mechanic arts, by which power, wealth, and political confequence, are in this age acquired. It exhibits a ftriking example of the evanefcent nature of wealth and power, when not fupported by induftry and economy. The feat of arts, of arms, and of Auguftus, after having been divided into a number of fmall principalities, of little confequence to the commerce or the politics of the reft of Europe, has at laft entirely loft its power and confequence, by the recent revolution effected there by the arms of France.

The induftrious manufacturers of the north will always be able to furnifh indolent Italy with fome of their manufactures, and will never ceafe to regain the productions of that luxurious and fertile country

Plate 18

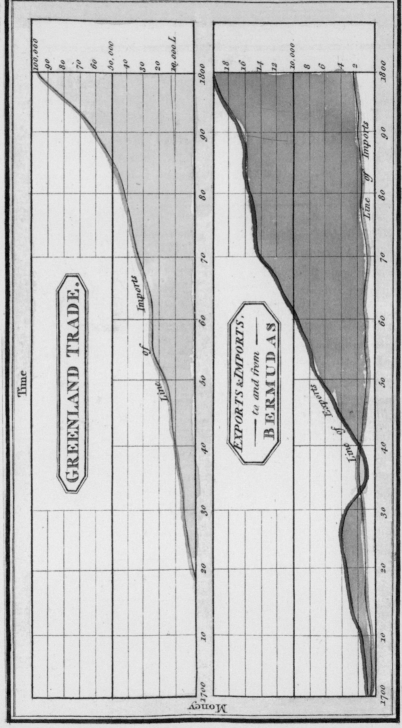

Time

Money

GREENLAND TRADE.

Line of Imports

EXPORTS & IMPORTS.
— to and from —
BERMUDAS

Line of Exports

Line of Imports

100,000
90
80
70
60
50,000
40
30
20
10,000

1800
90
80
70
60
50
40
30
20
10
1700

18
16
14
12
10,000
8
6
4
2

Neele sc. Strand.

OBSERVATIONS

ON THE

GREENLAND TRADE.

CHART XVIII.

THIS branch of our trade, which confifts of fifhing, (chiefly for whales,) is one, under circumftances peculiar to itfelf, and widely differing from thofe others on which we have been beftowing our attention.

In the divifion that takes place between mercantile and manufacturing bufinefs, fifhing ought to take its place under that of the latter: it is the production of manual labour, and not the transferring of property. If goods are carried from Surinam or Japan, their value is *but* increafed by change of place or change of poffeffor; but when a fifh is brought from the bottom of the ocean, its value, whatever it may be, is created; for no man would give a farthing for a fifh, even if it were a whale, while it is at large in the ocean. The chance of being able to catch the fifh is indeed of fome value; but that is not the value of the fifh, which depends entirely on its being caught. It may appear that paying rent for a fifhery, (of falmon for inftance) is giving money for the fifh that are in the water; but it is not fo in any degree; it is only giving money for the chance of catching the fifh. If the privilege of fifhing on a certain fpot belongs to an individual; or, in other words, if the laws of the country give an individual a right to all the fifh that are caught on that fpot, then may that right,

like

like any other, be let or fold, and the value of the fish is increafed by that price, in the fame manner as the price of corn is increafed by the rent of land. Nobody, however, would venture to fay, that the farmer who rented an eftate, paid the landlord for the corn that was to grow upon it; he would only fay, that he bargained for leave to cultivate the ground for his own advantage.

When there is no rent paid, the value of the fish confifts, altogether, in the labour and expenfe of catching them; and the effects, in a commercial view, are the fame as if a number of men created the fish out of nothing, or out of water, by the work of their hands.

Salt is univerfally faid to be manufactured, though the manner of procuring it is of the fame nature with fifhing, and, indeed, is performed in a manner not very unlike that in which fome favages catch fish; and though our procefs of making falt is, in its practice, not very like that of fifhing, in principle they are exactly the fame. The operation is, to feparate the minute particles of falt, which float in water, from the water itfelf. To do this, the water and falt are both put in a pan, and the water is evaporated by fire, leaving the falt behind. To feparate fmall fifhes from water, they are both put in a net-bag, and the water runs off through holes that are too minute to admit the fish to pafs, which are therefore left behind. So there is in reality a greater fimilarity between making falt and catching fmall fifhes, than there is between the fifhing for oyfters and for whales.

It may feem unneceffary to prove that fifhing is a manufacture, or (not to offend language) of the nature of a manufacture; but it is of high importance, that it fhould be generally

rally known to be fo; for manufactures, in point of doing good to a country, hold a much higher rank than merchandize, and are generally underftood to do fo. The exiftence too, of this branch of fifhery depends more upon public fupport and opinion, than many others; and it is one of the moft beneficial : therefore, whatever can tend to raifing its confequence, in a public view, is not only ufeful but neceffary.

The moft magnificent ftructures that art and labour have produced, which float upon the furface of the ocean, owe their origin to the defire that pervades equally the favage barbarian, and the moft luxurious prince, from the rude Hottentot to the moft fenfual Apicius, that of catching or of eating fifh. When that art was firft difcovered, neither philofophy nor public fpectacle had attained the pitch which is neceffary to induce a few individuals of the prefent age to venture in a new element; real neceffity, or fenfual appetite, were alone motives fufficient to attempt the unambitious natives of the dry land to venture upon the water : and inconfiderable as the portion of our prefent naval fkill required for fifhing may be, yet without the affiftance derived from that art, we cannot trace the different fteps by which we arrived at the prefent excellence in naval affairs. It has often been afferted, that fifheries are neceffary to maintain the ftrength and dignity of the Britifh navy, and it is unfortunate that it is not more univerfally believed.

Except agriculture and fifhing, the far greater number of the arts of peace tend to the effiminacy of the human frame. The other arts are inimical to that robuft body, and hardy mind, which will always be neceffary to the welfare of a ftate, while the nations fettle their difputes by force; and of thofe two, fifhing and agriculture, the former is the moft immediately

ately ferviceable to our ftrength in time of war; for in addition to ftrength of body and habits of life, it adds a degree of fkill that is capable of being foon made fufficient for the fervice of this country. Of all the different branches of the fifhery bufinefs, that carried on in the North feas for whales is perhaps the beft for a nurfery of feamen, and gives them a greater degree of health and ftrength than perhaps any other trade. The cold pure air, and the eafe they enjoy on board veffels that are pretty large and commodious, all together, make the men remarkably vigorous and healthy.

Befides the *real* advantage the fifhery gives, in raifing men for the fervice in time of war, it affords a *relative* advantage alfo, by preventing the Dutch, and other nations, from having a greater number of fifhermen than they at prefent have; for were it not for our rivalling the Dutch in fifheries in times of peace, we fhould not be able to rival them in fighting in times of war. Advantageous as this bufinefs is in all thefe refpects, in a ftrictly commercial fenfe it has not anfwered well yet; for it has never paid merely of itfelf, but has been fupported very wifely by a bounty from government. Could that bounty be increafed fo as to double the trade, it would be ftill better for the public.

The fifheries deferve encouragement on a greater number of claims than any other thing whatever; they will in the end be as productive of private wealth as they always have been, and will be productive of public ftrength, and national refpectability.

All thofe things that make a nation richer, ftronger, or more happy; or that tend to exalt national character, but that will not pay individuals, deferve public encouragement.

Learning

Learning and the fine arts exalt national character, and increafe happinefs; therefore, they have by great and wife monarchs generally been confidered as proper objects for patronage; for, without they were at firft taken by the hand, they would not pay the individuals, and therefore they would not exift; for it is an axiom, that what will not pay, will never exift in any extenfive degree; every thing muft either pay or be paid.

The veffels that are ufed for the Greenland fifhery are of a larger fize than coafting veffels, and it employs many veffels that would otherwife be ufelefs at the end of a war; by which means a confiderable lofs is prevented.

The bufinefs is a very precarious, rather than an unprofitable one, for it pays very highly fome years; but as the time in which the bufinefs is generally done, or loft, for the feafon, is but a few weeks (for before and after the time that the main body of the whales pafs, there is but little done) it is not expected to be very uniformly equal.

It appears from this chart that the trade increafes; but the real amount of bufinefs done is certainly confiderably above what it is here ftated to be, for this is not one of the articles that attracts much the minute attention of the cuftom-houfe officers.

With regard to the fmall trade to Bermuda, no obfervations occur.

FINANCIAL

FINANCIAL PART

OF THE

POLITICAL ATLAS.

OBSERVATIONS on the REVENUES

OF

ENGLAND and FRANCE,

FROM THE

Middle of the 16th Century till the prefent Time.

CHART XIX.

As England and France are the two powers in Europe which have ftood the moft prominently forward in deciding its differences, I have thought it would be well to give the progrefs of the revenues of each, in order to compare them from the earlieft period, that the documents are to be found correct, that is, from the year 1550.

Till the revolution in 1688, the Englifh line of revenue cannot be abfolutely depended upon, but it is not far wrong. From that epoch, when our national debt firft begun, the expenfes have rifen with an immence rapidity, and it is eafy to fee that they muft continue to do fo. Not only do the expenfes increafe as the intereft of debt augments, but the free revenue neceffarily keeping pace with the expenfes of the times, adds confiderably to our annual embarraffments by augmenting alfo.

G.   At

At this moment, were we to make peace, the peace eftablifhment could not probably be lefs than 30,000,000; though,
previous to this war, as we fee, it was only 16 millions, which
is a fudden change indeed, and one that will be the more
felt, that France, with whofe expenditure was heretofore
greater than our's, will now be not equal to one-half of it.——
This work is intended to reprefent the paft, and by no
means lead to obfervations on its future increafe. Each
perfon muft then draw his own conclufions from the facts
here exhibited in fo legible a character. Equals in bravery, and emulating each other for ages in the arts of
peace, and the exertions of war, fometimes friends and fometimes enemies, but always rivals, we fee the two nations
advancing in revenue and expenditure with a pretty equal
pace, till the revolution of 1789 changed the bufinefs, and
has very nearly freed France from all her debts, while it has
almoft doubled thofe of England. So great a change muft
have important confequences. The equilibrium, that hitherto
exifted, is certainly deftroyed; it is indeed true that France
has confumed its capital in a great meafure, as well as got
rid of its debts,* but capital can never be long wanting in a
nation fo fertile, and fo full of induftrious and active inhabitants as France. The rapidity with which public affairs approach a crifis is fuch as promifes a fpeedy folution of the
queftion, as to the effects of fo fudden a change on the
finances of two rivals and neighbours, whofe increafing expenfes have kept pace with each other for nearly three centuries.

* France paid about the fame fum for intereft of money borrowed that
England did before the year 1789; but one half was on annuities (*Rentes
Viageres*), fo that the capital owing was not equal by above three-fourths.

OBSER-

and of Britain would never have reimburſed the original ex‑
penſe of conqueſt.   It has only been in the modern times,
that war has become the buſineſs of the ſtate alone, and not
of individuals.   During the preſent conteſt, the French have
adopted the ancient ſyſtem with ſucceſs.

It is not neceſſary to inſiſt on the impoſſibility of laying up
money to ſerve for future wars, as it is totally inconſiſtent with
the preſent ſtate of things ; the mode of borrowing, from
money holders, is certainly much better adapted to the cir‑
cumſtances of a commercial ſtate ; and the principle upon
which the Britiſh funds are conducted is a great improvement
upon the original mode of borrowing ; but, like other things
that anſwer well, we are too ready to have recourſe to it.

There are only two ways of borrowing ; the one is at the
natural rate of intereſt of the country, in which caſe the ca‑
pital remains unpaid, and ſtill due to the lender ; the other is,
when more than common intereſt is paid ; in which latter caſe,
both capital and intereſt determine and end at a certain day.
All loans, whatever may be their particular nature, come
under one or other of theſe heads.   There are likewiſe only
two motives for borrowing money.   One is to uſe in trade,
or to improve grounds, in which caſe a gain is generally pro‑
duced that pays the intereſt as it becomes due, beſides a pro‑
fit, which, in time, repays the original debt, and leaves ſome
gain remaining to the borrower.

The other occaſion for borrowing money is merely to uſe
in living, in expenſe, in pleaſure, or to pay debts ; in which
caſe no profit is produced, and the borrower muſt pay both
the intereſt and capital from other funds or reſources, as itſelf
produces none.

This

This laft-mentioned caufe for borrowing is ruinous in its nature, as the other, when wifely done, is advantageous; and if loans of this fort are to be contracted, it is moft advantageous to borrow them at a high rate of intereft, to determine at a certain day. Money borrowed for the purpofe of carrying on bufinefs, on the contrary, fhould be borrowed upon the loweft rate of intereft, as it is producing a fund to repay the capital.

The nature of national loans is of the firft of thefe kinds that do not produce a fund; therefore they fhould be at high intereft, that they may determine at a certain day; for if not, they will continue to accumulate.

The effect, which is before us in the chart, is the natural one of perpetual loans; for though it might have been managed a little better, or a little worfe, as long as it continued upon this plan, it muft have increafed, unlefs as much money had been levied in time of peace, as would pay off the debt contracted in the previous war; but if this were done, it would be the fame with annuities, becaufe it would be raifing taxes to pay off, at a certain time, the capital of the debt. That, however, has not been done, except, during this laft war, content with getting poffeffion of the money, we have left to future generations the trouble of repaying it.

The national debt, as here reprefented, is in real millions fterling, not in the nominal millions of three per cents. four and five per cent. ftock. I have confidered it as all borrowed at five per cent. and as being at par, that is to fay, I eftimate the capital by the annual intereft, multiplying the latter by 20, this mode is, I think, more accurate than any other, and is infinitely more fimple.

G 3

In

In the year 1786, a finking fund of an annual million was eftablifhed for the paying off the capital of the debt which, if ftocks were always at par, it would accomplifh in about 53 years, and, fince that period, no new loan has been made that has not had a finking fund attached to it, nearly in an equal proportion, that is to fay, of one per cent. annually on the capital, the effect of one per cent. will be to reduce the capital in about 50 years alfo, fo that we may now confider the whole of the debt as reducible in nearly about that term.

So many calculations and theories have been made and entered into about the cataftrophe that the accumulated debt would produce, and the fum it now amounts to is fo much greater than was thought to be poffible, that people are now become totally incredulous with refpect to the crifis that evidently approaches, we fhall therefore fay nothing on that fubject. One party fuppofes the evil nearly at hand, another difbelieves every prediction of the fort ; and the truth is, that from the diminution, or depreciation of the value of money, added to the increafing refources of the nation, our load of debt lies much lefs heavy on us than our greateft calculators expected ; but, of late years, the effects they foretold are becoming manifeft. The high prices of provifions, the demands made every where for the increafe of the price of labour, and the general penury of the middle and lower claffes—thefe are the effects of heavy taxation, of which our debts alone are the caufe.

OBSER-

Plate 21

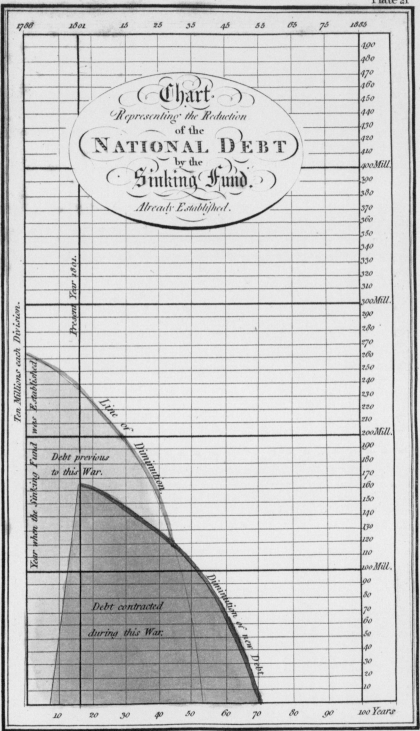

Chart.

Representing the Reduction

of the

NATIONAL DEBT

by the

Sinking Fund.

Already Established.

1786 1801 15 25 35 45 55 65 75 1865

490
480
470
460
450
440
430
420
410
400Mill.
390
380
370
360
350
340
330
320
310
300Mill.
290
280
270
260
250
240
230
220
210
200Mill.
190
180
170
160
150
140
130
120
110
100Mill.
90
80
70
60
50
40
30
20
10
100 Years

Ten Millions each Division.

Present Year 1801.

Year when the Sinking Fund was Established.

Line of Diminution

Debt previous to this War.

Debt contracted during this War.

Diminution of new Debt

10 20 30 40 50 60 70 80 90

Neele sc.Strand.

OBSERVATIONS ON THE OPERATION

OF A

SINKING FUND OF ONE MILLION,

Applied unalienably to the Reduction of the National Debt.

CHART XXI.

THIS chart is of the same dimensions, and upon the same scale with that which represents the increase of the debt, in order that no mistake or perplexity may arise in drawing between the two an exact comparison.

The part that is stained blue represents the manner in which the debt will diminish. The line of years, from which the numbers at the bottom begin, is the year on which the operation of the fund commenced. The curved line by which the blue part is bounded, is that in which the debt will decrease. This curve, reckoning the probable expense of management, and the inconsiderable annuities that will fall in, is very near the calculation; and the most pertinacious cannot say that it differs two years from that on which the fund was instituted. Every calculation, respecting the payment of the debt, must in some degree be founded upon uncertainty, as the price of the funds at the time of redemption will accelerate or retard the progress in proportion as they are above or below par, and there is no possibility of calculating what those prices may be at any given time.

G 4

The

The curved line, which is nearly a parabola, shews the pro-gressive diminution of the debt, previous to the loans which have been made during this war; and to each of which a sinking fund of one *per cent.* has very properly attached, by which the money borrowed will be paid off, as shewn by the yellow line.

The accelerating progress of a fund, acting on the princi-ple of compound interest, has been thrown as it were into ridi-cule, by the extravagant calculations of arithmetical pedants; not that they have made wrong statements, but by imagining cases, which the nature of things prevent from ever being realized. The accumulation of one penny, for a great number of years, producing a sum which would pay the national debt, though in theory true, is so extravagant, that one would think it was stated in order to give the appearance of fiction to what is the result of accurate calculation.

The case in fact is, that when money has accumulated to a considerable sum, it becomes impracticable to employ it with-out expense and risk and both the risk and expense increase in a double proportion with the amount of capital; and finally, the employment of the money would become impracticable. The theories, therefore, of accumulation appear fictitious; and they are so to a certain degree, that is, that they cannot be realized.

In regard to paying off debt the case is quite different; for so far as that goes, the full advantage is derived, and the pro-gress still farther hastened by the diminution of expenses on the administration. Those theories that are incapable of being realized in accumulating millions, are quite practicable in pay-ing them off, a truth, of which England already begins to feel the good effects.

OBSER-

Plate 22.

Chart of Money Voted for SERVICES since the Year 1700.

Scale as Pound.

OBSERVATIONS

ON THE

MONEY GRANTED FOR SERVICES,

From the Year 1722 *to the Year* 1800.

CHART XXII.

THIS is taken from the accounts laid before parliament of the services of the different years. The amount depends chiefly upon whether it is war or peace at the time.

These grants are the original form in which the national debt existed. The manner is shortly thus: Troops, ships, &c. are first voted by parliament, for the service of the year, and estimates made from the votes, are laid before the house, and whatever they amount to more than the receipts of the treasury, is borrowed; the parliament laying on taxes to pay the interest. If the taxes produce less than that interest, the deficiency is supplied by additional ones next year; and if they amount to more, the surplus goes in diminution of the loan of the following year; or if it is time of peace, and no loan is wanted, it is applied to paying off debt.

The amount of supplies is, upon the whole, on the increase; for though they have fallen at the end of each war, yet they do not return to their usual state.

The

The value of money is not the fame that it was in the year 1720: this is one reafon for part of that increafe ; but, it evidently appears, that expenditure in every line rifes in a proportion greatly exceeding the depreciation of money, it would however exceed the bounds of this work, and would be a deviation from the plan, were we to attempt to enter into an examination of the various caufes, of which, however, the pepreciation of money muft be allowed to be one of the greateft.

OBSER-

Plate 25.

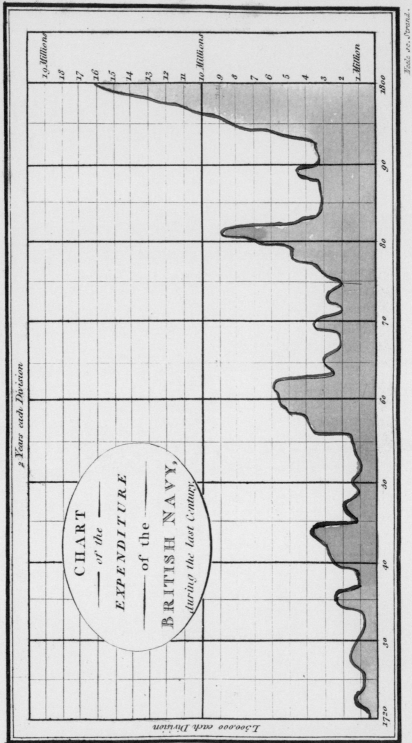

OBSERVATIONS

ON THE

EXPENDITURE of the NAVY.

CHART XXIII.

THIS department confumes a confiderable part of the money granted for *fervices*, and rifes the fame years that it rifes, though not exactly in the fame proportion. This alfo is taken from the accounts of eftimates laid before parliament. From the nature of the navy, the extraordinaries, as they are called, muft vary much in different years. The expenfe of building fhips is great, and their fate, when built, uncertain; which has occafioned an impoffibility of eftimating the expenfe before hand; but paying in navy-bills prevents that neceffity, and alfo prevents parliament from having that controul over the difburfements that it has over the other branches of public expenditure.

Although the brilliant atchievements of our brave admirals, officers, and feamen, is not a theme for expatiating upon in a book on the finance of the country, yet it is fcarcely poffible, in laying before the public the expenditure of that department, not to fay fomething on a fubject fo gratifying to Englifh minds, and in itfelf fo unexampled.

The Roman arms, in the beft times of the republic, were never more univerfally and uninterruptedly fuccefsful on land than our fleets have lately been by fea. Whatever has been the

position

pofition or the force of the enemy's fhips, ours have never, either individually or in a fquadron, attacked without gaining the victory, fome of which have been of the moft fplendid fort, and all of them honourable for the officers and men engaged.

Whatever, in the event of things and in the courfe of time, may be the fate of Britain and the nations with which fhe has contended fo bravely by fea, the atchievements of the Britifh navy, during the prefent war, will throw a luftre on its hiftory that will intereft future ages, as the brilliant actions of the Greeks and Romans have done all thofe who have followed them, and as they ever will excite a lively intereft.

It will fcarcely efcape the obfervation of thofe who attentively confider this chart and thofe for the army and ordnance, that the laft years of war are very expenfive, and the two or three firft coft proportionally but little, and that it has uniformly been fo, from which the evident and direct conclufion is, that our efforts, at the beginning of a war, are feeble, and that the national exertions are only called forth by degrees, than which, no mode of acting can be more ruinous. It is well known that the American war was badly conducted in this refpect, and the prefent one has not been much better, for though bravery has triumphed, yet inferiority in numbers has coft much blood and hard fighting which might have been faved.

The duration of wars would be leffened and fuccefs infured, as far as fuch a thing is poffible, were this nation, the moment it began a quarrel, to difplay its whole force and vigour, and ftrike fome great blow before the enemy could be prepared. It is to be hoped that in future this will be the cafe.

OBSER-

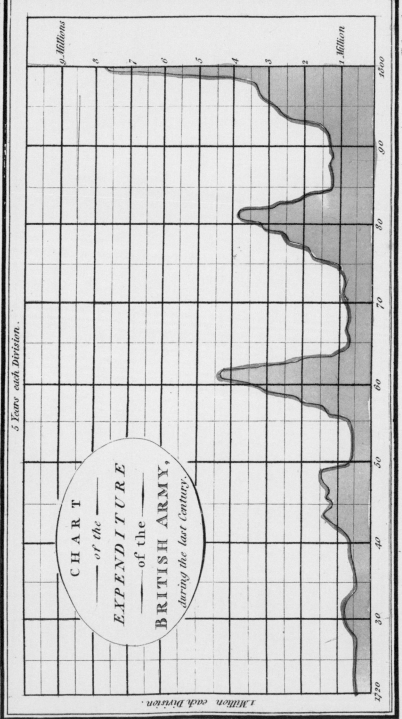

5 Years each Division.

9 Millions

8

7

6

5

4

3

2

1 Million

1 Million each Division.

CHART
— of the —
EXPENDITURE
— of the —
BRITISH ARMY,
during the last Century.

1720   30   40   50   60   70   80   90   1800

Scale & Sowerby.

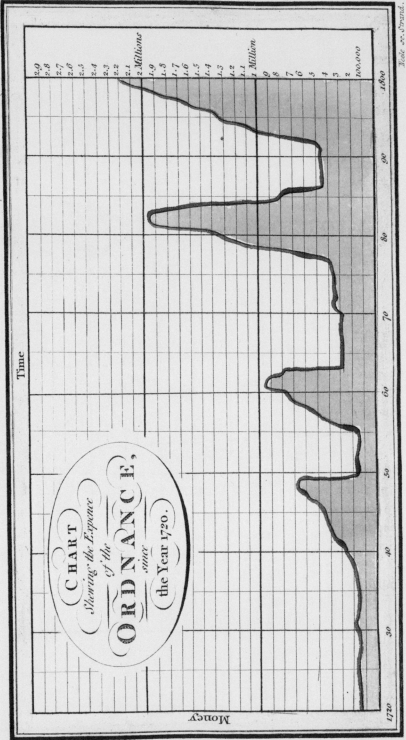

Plate 25

Time

Money

2.9
2.8
2.7
2.6
2.5
2.4
2.3
2.2
2.1
2 Millions
1.9
1.8
1.7
1.6
1.5
1.4
1.3
1.2
1.1
1 Million
.9
.8
.7
.6
.5
.4
.3
.2
100,000

1720 30 40 50 60 70 80 90 1800

CHART
Shewing the Expence
of the
ORDNANCE,
since
the Year 1720.

Vide St. Strand.

OBSERVATIONS

ON THE

EXPENDITURE ON ACCOUNT

OF

LAND FORCES AND OF THE ORDNANCE.

CHARTS XXIV. AND XXV.

THESE chart rifes and fall like that of the navy, only more
fuddenly. It would be very defirable to know the numbers
of troops retained in pay at different times; but the eftimates
and papers laid before the houfe, do not ftate it with fufficient
accuracy, to enable us to do that.

The expenditures of the army feem to have been very great,
when we confider the number of men, and their pay. Sup-
pofing there were 100,000 men employed, at 4s. a day, one
with another, officers and men, it would not have amounted
to the expenfe, either in Germany, or during the laft war in
America. But the different fums expended for hired troops
from Germany, &c. &c. make it very difficult to compare the
number of men with the money expended.

The pay of foldiers is lately raifed, and many other circumftances tend to increafe expenfes of the late years.

The

The obfervations that apply to the land army in general, apply likewife to the ordnance, which, properly fpeaking, is only a particular branch of that department. Our numerous colonies and garrifons abroad have had a great fhare in augmenting the expenfes of the ordnance.

OBSER-

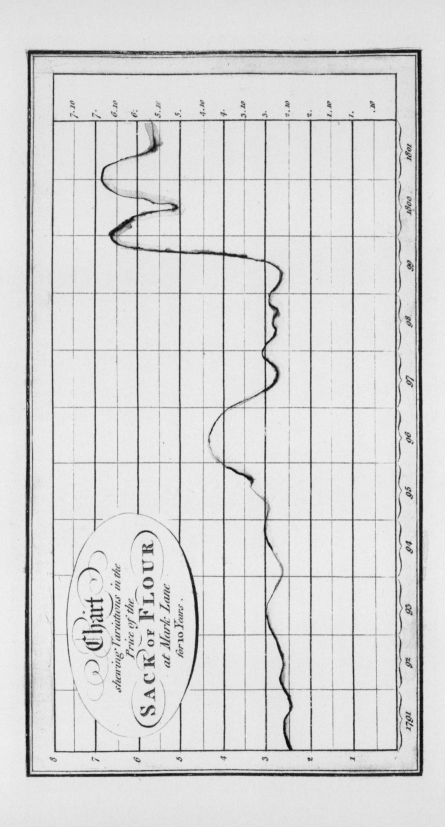

Chart
shewing Variations in the
Price of the
SACK of FLOUR
at Mark Lane
for 10 Years.

OBSERVATIONS on the PRICE

OF

FLOUR during the last TEN YEARS at MARK-LANE.

CHART XXVI.

THIS fubject, in itfelf fo important, has of late attracted the public atten-tion in a very high degree. Thofe who are for the liberty of commerce, (an improper phrafe) wifh the buyer and confumer to be left equally free. Another clafs of perfons recommend making laws and regulations to enforce fales of corn under certain prices. The queftion is as intricate as it is important, but too long to difcufs here, although it might be wrong to pafs it over entirely in filence.

Perfect freedom of commerce, that is to fay, commerce without regula-tions, is juft as wild an idea as the perfect liberty and equality conceived by the French in 1789; and it is not a little fingular that both opinions were foftered and fupported by the fame fect of economifts. Law muft, in fome cafes, interfere to counteract the machinations and combinations of men; the queftion then is, not a pure queftion to be difcuffed by reafon, but one, in the folution of which, policy and experience muft enter and operate.

The commerce carried on in grain is already under fome reftriction; therefore, in fact, the queftion of the interference of law and regulation is determined already, and it only remains to find out how far and in what cafes it ought to be applied.

The great and well-deferved reputation of Mr. Adam Smith, one of the ableft writers on commerce, who was an advocate for free trade, on the principle *that free trade finds its level*, has made many people blindly follow his principles, and, to quote his opinion, has ftood in place of argument; but with all due deference to the opinion of that great man, (with whom the author of this had the honour of being perfonally acquainted,) it muft

3   be

be allowed that he in this argument omitted to divide commerce into two sorts, which are governed by different principles, his arguments and theories being all applicable only to one species of commerce, that is, where the articles are such that the seller and the purchaser are entirely independent of each other.

When the demand for an article is dependent on taste, caprice, or the will of the buyer, he is totally free from the seller; where the article is necessary, but not so in any immediate manner, or at a fixed moment, the consumers enjoy a degree of independence, that prevents the necessity of any interference of law or regulation; but where the quantity consumed is regulated by the imperious calls of nature, and even the hour and minute, as it were, fixed for supplying them, the consumers become dependent on the sellers. This is evident; but then say the advocates for freedom of trade, the concurrence of sellers will regulate that: if, however, it should be found, that, instead of underselling each other, the sellers should either by seeing the matter in one light, or by any cause whatever, all coincide in keeping up the prices, then the second argument falls to the ground: now, it is a fact, that the venders of grain have united in asking great prices, so that some regulation is become necessary.

To give an example:—Toys and trinkets may be sold at what price the sellers may ask; it would be absurd to attempt any regulation, and it never was thought of in any country. The wages of certain sorts of labour, and the price of fuel and beer have often been fixed because combinations were possible. This proves that there are two branches of commerce, fundamentally different, and therefore not to be reasoned upon precisely in the same manner. Mr. Smith, therefore, has fallen into an error, and never could have literally meant that he expressed, as in other parts of his work, particularly on the interest of money, he has recommended regulation.

This chart is meant, like the others, to shew the fact in a strong point of view, that those who see it may reason from it; and it is certain, that if the price of grain does not *soon* fall, the rent of land and wages of labour will rise, so as to prevent the grain from ever falling again to near its former price, under the present order of things. This would aim a blow at our commercial relations with the rest of Europe, that might destroy the whole system on which the prosperity of this country is founded; therefore, if the cries of the hungry and the needy will not make those in power interfere, let them pay attention to the dearest interests of their country.

THE END.

T. BURTON, PRINTER, LITTLE QUEEN-STREET.

TABLE of CONTENTS.

FINANCIAL PART.

PREFACE.

HAVING about a year ago been requested by the English editor of Mr. Boetticher's Statistical Tables, to consider of some method of bringing them down to that period, without injuring the original work, I proposed to make a *supplementary* table, comprehending all the countries which have undergone any material change since the publication of the book. I then undertook to make out such a supplementary table; which I did, and it is published at the end of that work.

In the course of executing that design, it occurred to me, that tables are by no means a good form for conveying such information, unless where a number of different countries are intended to be exhibited at once. Where there is only one to be set forth, I can see no kind of advantage in that sort of representation, while the inconveniency of a large size, in a book that is intended to be frequently referred to, is obvious.

A 2   I do

I do not conceive that it is in any manner the province of ſtatiſtical works to contain hiſtorical relation, or any thing that is not a ſimple fact, and relative to one ſingle epoch or date. The numbers of people, quantity of ground, revenues, prices of labour, &c. as ſimple and uſeful facts, belong to ſtatiſtics; but the deſcription of the order of the garter, or of the golden fleece, has nothing to do with it. To encumber ſtatiſtical reports with ſuch information appears to me to be ill placed, and as ſuch improper.

I have compoſed the following work upon the principle of which I ſpeak; this, however, I never ſhould have thought of doing, had it not occurred to me, that making an appeal to the eye when proportion and magnitude are concerned, is the beſt and readieſt method of conveying a diſtinct idea.

Statiſtical knowledge, though in ſome degree ſearched after in the moſt early ages of the world, has not, till within theſe laſt fifty years, become a regular object of ſtudy. Its utility to all perſons connected in any way with public affairs, is evident: and indeed it is no leſs evident that every one who aſpires at the character of a well-informed man ſhould acquire a certain degree of knowledge on a ſubject ſo univerſally important, and ſo generally canvaſſed.

Geographical knowledge has long been conſidered as neceſſary neceffary for perfons of both fexes who wifh to acquire any tolerable degree of general information ; in fo much that, next to ignorance of the grammar of one's native language, nothing betrays want of information fo foon as ignorance in matters of geography, without which it is almoft impoffible to carry on converfation long on any general fubject.

Geography is, however, only a branch of ftatiftics, a.knowledge of which is neceffary to the well under-ftanding of the hiftory of nations, as well as their fitua-tions relatively to each other. In ancient hiftory, and even down to our own times, there is nothing fo imper-fect as the accounts given of ftatiftical matters. Ancient hiftorians, and other writers, tell us for example, of great armies raifed and great achievements performed ; but concerning the finances, and ways and means, they are generally filent. To the importance of this fpecies of knowledge, mankind have only of late years begun to pay a fufficient degree of attention, the want of which, hitherto, leaves us now in great ignorance on many points which it would be very ufeful for us to know, in, order to form a comparifon between the ancient ftate of the world and its prefent fituation.

Statiftical accounts are to be referred to as a dictionary by men of riper years, and by young men as a grammar, to teach them the relations and proportions of different

ftatiftical

ftatiftical fubjects, and to imprint them on the mind at
a time when the memory is capable of being impreffed in
a lafting and durable manner, thereby laying the foundation for accurate and valuable knowledge.

Since the value of this ftudy is generally acknowledged,
it has become a defirable thing to render it as eafy
and perfect as may be.　In the introduction reafons
are given for adopting the mode of reprefenting the
magnitude of different countries by proportional circles,
but the great teft of its utility is in the mind of the perfon who takes up the charts.　The firft of thefe has been
fhewn to numbers of perfons, all of whom have declared
that till they faw it, they had no right and diftinct idea
of the proportional extent of the different countries fuch
as it gave them.　The reafon of this is evident: for, as it
is not without fome pains and labour that the memory
is impreffed with the proportion between different quantities expreffed in words or figures, many perfons never
take that trouble ;—and there is even, to thofe that do fo,
a frefh effort of memory neceffary each time the queftion
occurs.　It is different with a chart, as the eye cannot
look on fimilar forms without involuntarily as it were comparing their magnitudes.　So that what in the ufual mode
was attended with fome difficulty, becomes not only eafy,
but as it were unavoidable.

Whatever prefents itfelf quickly and clearly to the

mind,

mind, fets it to work, to reafon, and think; whereas, it often happens, that in learning a number of detached facts, the mind is merely paffive, and makes no effort further than an attempt to retain fuch knowledge.

It would be almoft impoffible for any perfon of intelligence to contemplate the firft chart without being ftruck with the great fize of Ruffia and Turkey, and the comparatively fmall extent of thofe countries which have borne the principal fway in the world for thefe laft five hundred years, whilft Ruffia was nearly unknown, and counted but as duft in the political balance of nations. Some general conclufions, accompanied with no fmall degree of furprife, naturally attend the firft view of this proportional chart of nations.

What thinking man who confiders the important part that the fmall republic of Holland has acted, while Ruffia lay as if congealed in an eternal winter, but will conclude, that if ever the people in thofe different countries come to be in any degree fimilar in civilization and intelligence, the importance of the fmaller muft fink into great inferiority, and in general, that if even the different countries in the world fhould come to be nearly upon a par in refpect of arts, civilization and knowledge, the fcale of their importance muft be ftrangely altered, and accordingly it is daily altering: for, as commerce, arts, and civilization, have been making great progrefs

A 4   during

during the laſt century, the foundation of changes has been ſolidly laid, and they have begun to take place with unexampled rapidity.

Holland, which was a preponderating power in the beginning, and during a great part of the laſt century, as it had long before been, entered into the laſt war ſhorn of its importance, with the rank of only an auxiliary to France and Spain. It did not long preſerve even that diminiſhed rank; for having firſt ſubmitted to be the tool of a French faction, it was in the courſe of a few days reduced to obedience by the King of Pruſſia, who acted with it juſt as he would have done with a re-bellious province of his own dominions; and when the preſent war broke out, it ſoon was reduced to what impartial truth obliges us to call a dependant province of France.

Portugal, now ſo different from what it was in the time when its conqueſts almoſt encircled, and did aſtoniſh the world, ſeems to run a riſque of ſharing the fate of Holland.

Though extent of territory is the ground work of power, as it regulates in a great degree the population of a country; yet we find neither extent nor population will do without revenue: hence we find Poland extenſive, populous, rich in ſoil, and productive, peopled with a race much more zealous of liberty than any of the neigh-

bouring

bouring kingdoms, fallen a prey to the power of thofe very neighbours. The conclufion is, that want of revenue was the caufe of its ruin*.

To render ftatiftical accounts accurate and complete, it is not fufficient that individuals fhould collect knowledge, and arrange it in order, for the aid of rulers and magiftrates. An habitual and regular practice of collecting information, both generally and locally, is neceffary; but as vanity is not flattered by employing men to collect fuch materials, as it does not immediately advance the interefts of thofe who are at the head of affairs, it is to be feared that the bufinefs will long be left to the inadequate care of a few individuals.

Where vanity is not gratified, or intereft promoted, knowledge is generally neglected. The bufhels of rings taken from the fingers of the flain at the battle of Cannæ, above two thoufand years ago, are recorded: fo are the numbers of combatants at the battles of Agincourt and Creffy; but the bufhels of corn produced in England at this day, or the number of the inhabitants of the country †, are unknown, at the very time that we are debating that

most

* Perhaps it will be urged that want of unanimity, not want of revenue, ruined Poland; but in anfwer to this, it may be urged, that want of revenue occafions want of unanimity as well as many other wants.

† Some efforts have been lately made to afcertain the population of this country, which are entirely inadequate to the purpofe, and are therefore to be confidered as nothing.

moft important queftion, whether or not there is fufficient
fubfiftence for thofe who live within the kingdom. We
neither know whether the country is increafing or diminifhing in population : we are equally ignorant as to
its produce, and yet, perhaps, no nation in Europe is
better informed on thofe important fubjects than our-
felves. No encouragement is given, no proper fteps are
taken by thofe who rule, to afcertain points that are fo
material, while there are Societies inftituted for inquiry
into matters which are paft and gone, rare and ufelefs,
or diftant and unknown.

Were the aid and fupport of public men obtained in
collecting ftatiftical knowledge, great progrefs might be
made in it at little expence, and with great facility ; but
fo long as that is not the cafe, individuals will find themfelves reduced to the fituation of fcanty gleaners, not
that of men carrying home an ample harveft.

Statefmen, and thofe in power, would in the end find
themfelves amply repaid for any trouble, or moderate degree of expence, that an attention to ftatiftics might occafion ; for by that means the operations of government
(particularly the revenue department *) would be greatly
faci-

---

* In the revenue department much accuracy and great attention prevails
throughout ; but all other national operations are done in a flovenly inaccurate manner, as if revenue alone were worth attending to : it is not fo in
many countries that are in other refpects much worfe regulated than this.

facilitated. Great ſtateſmen and monarchs have known
this in all ages; from whence attempts have ariſen to
number the people, and take an account of property, &c.

As ſtatiſtical reſults never can be made out with minute accuracy, and that, if they were, it would add little
to their utility, from the changes that are perpetually
taking place; it has been thought proper in this work
to omit that cuſtomary oſtentation of inſerting what may
be termed fractional parts, in calculating great numbers,
as they only confuſe the mind and are in themſelves an
abſurdity.

Statiſtical books, like dictionaries, require new editions
from time to time, as changes take place among nations;
but it is impoſſible to begin a regular ſeries of ſuch accounts from any period ſo proper as that juſt previous to
the preſent war. Europe had been almoſt ſtationary for
a century, when all at once changes commenced, which,
from their nature, their cauſes, and the general ſituation
of things, will not ſoon be ended in a ſolid manner. The
firſt view of European nations is the foundation from
which we riſe, with an intention to exhibit in a like manner the ſame nations under the different viciſſitudes which
the preſent troubles have occaſioned, or in future may
occaſion.

ADVERTISEMENT.

In the obfervations made relative to the utility and fitnefs of large tables for conveying ftatiftical information, no idea was entertained of objecting to the merit of M. Boetticher's work ; but from infpecting thofe tables themfelves, it will appear, that except in regard to Germany, which is divided into a great number of governments, in the VIth table, where eight ftates are reprefented at once, and in the laft fupplementary table, where eleven different nations are contained, there is more inconveniency than advantage arises from the form adopted.

With refpect to throwing afide the units, tens, and hundreds, in great numbers, it is done under this fimple impreffion, that as the information does fcarcely ever come within a thoufand of the truth, it is an affectation of accuracy beyond what has really been attained; or, to to make a fair comparifon, it is like a hiftorian giving as truth, an account of the private minutiæ of courts and embaffies, which were known only to the parties themfelves, and though reported publicly never believed. No fort of reflection is however meant on thofe who think fit to give their ftatements in the other way, although the number of figures certainly embarraffes the memory without anfwering any good purpofe.

INTRODUCTION and EXPLANATION

OF THE

STATISTICAL CHARTS.

———

Each circular figure reprefents that country, the name of which is engraved under it, and all are arranged in order according to their extent.

The lines ftained red that rife on the left of each country, exprefs the number of inhabitants in millions, meafured upon the divided fcale which extends from right to left of each divifion, each of which is one million, as marked at both ends.

The yellow lines on the right of each nation reprefent the revenue in millions of pounds fterling, meafured alfo upon the fame divided fcale with the population.

The countries ftained green are maritime powers; thofe ftained of a pale red are only powerful by land.

The figures marked directly above the circles (as 5 over Ruffia, and 14 over Sweden) indicate the number of perfons living on each fquare mile of country.

The figures within the circles fhew the number of fquare miles in the countries they reprefent.

The dotted lines drawn between the population and revenue, are merely intended to connect together the lines

belonging

belonging to the fame country. The afcent of thofe lines being from right to left, or from left to right, fhews whether in proportion to its population the country is burdened with heavy taxes or otherwife.

CHART 1ft. Exhibits the powers of Europe as they were previous to the French Revolution.

CHART 2d. The nations of Europe, as intended by the peace figned at Luneville, which fo materially alters the nature of affairs, and the extent of France and Germany,

CHART 3d. Reprefents the population of the great capital cities of Europe, the circles being proportioned to the number of inhabitants in each.

CHART 4th. Reprefents the powers of Hindooftan, that are connected with, or influence European affairs in the eaft, in the fame manner that the European powers themfelves are exhibited to view.

The advantages propofed by this mode of reprefentation, are to facilitate the attainment of information, and aid the memory in retaining it : which two points form the principal bufinefs in what we call learning, or the acquifition of knowledge.

Of all the fenfes, the eye gives the livelieft and moft accurate idea of whatever is fufceptible of being reprefented to it; and when proportion between different quanties is the object, then the eye has an incalculable fuperiority; as from the conftant, and even involuntary habit of comparing the fizes of objects it has acquired the capacity of doing fo, with an accuracy that is almoft unequalled.

The

The ftudy of chronology has been much facilitated by making fpace reprefent time, and a line of a proportional length, and in a fuitable pofition, the life of a man, by means of which the remarkable men of paft ages appear as it were before us in their proper time and place.

The author of this work applied the ufe of lines to matters of commerce and finance about fixteen years ago, with great fuccefs *. His mode was generally approved of as not only facilitating, but rendering thofe ftudies more clear, and retained more eafily by the memory.

The prefent charts are in like manner intended to aid ftatiftical ftudies, by fhewing to the eye the fizes of different countries reprefented by fimilar forms, for where forms are not fimilar, the eye cannot compare them eafily nor accurately. From this circumftance it happens, that we have a more accurate idea of the fizes of the planets, which are fpheres, than of the nations of Europe which we fee on the maps, all of which are irregular forms in themfelves as well as unlike to each other.

SIZE, POPULATION, and REVENUE, are the three principal objects of attention upon the general fcale of ftatiftical ftudies, whether we are actuated by curiofity or intereft; I have therefore reprefented thefe three objects in one view, as they are the only effential foundations for power that can be accurately meafured or laid down with mathematical precifion. Forms of govern-

---

* In the Political and Commercial Atlas, delineating the progrefs of the commerce and revenues of this country during the laft century. That work was tranflated into French and publifhed in Paris in 1788, when it met with great approbation. A new edition up to the prefent time is juft publifhed, of a fize to bind up with this.

ment

ment, and the moral qualities of men, go a great way in conftituting the ftrength of nations ; but thofe can only be defcribed to the mind, they cannot be reprefented, nor indeed are they even fufceptible of accurate defcription.

To conclude, the 1ft chart fhews the different powers of Europe at one view ; by which the mind may conceive, and the memory retain, a diftinct idea of their proportional extent, population, and public revenues. As for the details of each individual nation, they are fimple, not comparative facts, and are to be found in the printed table dedicated to that particular country.

Thofe who will take the trouble to read the preface, will find in it fome other remarks on this new mode; which may deferve their attention ; but, as already obferved in that preface, the great criterion is the effect produced on the mind of a man, when it has for a few minutes contemplated one of thefe charts for the firft time.

It is prefumed that to ftudents this work will be particularly ufeful: for no ftudy is lefs alluring or more dry and tedious than ftatiftics, unlefs the mind and imagination are fet to work, or that the perfon ftudying is particularly interefted in the fubject; which laft can feldom be the cafe with young men in any rank of life.

N. B. Should future changes require a new chart, it will be publifhed of a fize proper to bind up with this work, and will be fold, to thofe who have a copy of it, at half the price charged to the public.

STATISTICAL ACCOUNT, &c.

THE TURKISH EMPIRE.

THE fineſt portion of the world is in poſſeſſion of the Turks, ſince the year 10 0. The government is deſpotic, with power over both the perſon: and property of the ſubject. There is a great difference between a deſpotic government in a Mahomedan and in a Chriſtian country,—in the former, it is not reſtrained by the tenets of religion ; whereas we have no inſtance of any Chriſtian king being guilty of ſuch acts of violence as are every day practiſed at the courts of Mahomedan princes.

This great Empire, next in magnitude to that of Ruſſia, and about equal to it in population and revenue, has undergone many revolutions, and is conſidered as on the decline for this laſt century. Certain it is, that it has loſt much of the energy it has on former occaſions diſplayed ; but that does not always mark decay in countries ſo governed, where the character and talents of thoſe who rule at the time, have a preponderating influence on public affairs.

The hiſtory of the Turkiſh Empire is too full of events to admit of any thing like an abridgment here ; but the Turks formerly made a tolerable equally-poiſed rivalſhip with the Germans by land, and with the Venetians by ſea. So late as 1789, Ruſſia and the Emperor united, were both kept at bay by Turkey, and one campaign was very brilliant; nevertheleſs, nothing has been more feeble than the efforts made by that power to co-operate with this country in Egypt, or to ſubdue Paſhwan Oglu. Caprice, or ſome cauſe, the real nature of which is little known, ſeems to produce alternate fits of exertion and of inactivity.

The great aggrandiſement and progreſſive improvement of the Ruſſian Empire, is indeed a dangerous circumſtance for Turkey ; but perhaps other European powers are not quite ſo loſt to all juſtice and to their own intereſts, as to look on with indifference at the ruin of ſo great an Empire.

Extent

Extent in square miles 790,000
Number of inhabitants 24,000,000
Number of persons to a square mile 31
Extent in English acres 505,600,000
Number of acres of land to one person 21
Revenue in pounds sterling 7,000,000
Amount of public debt, none
Land forces in time of peace 250,000
Ditto, in time of war 370,000
Seamen in time of peace 30,000
Ditto, in time of war 55,000
Ships of the line 40
Frigates, sloops, &c.—20 frigates, 40 gallies, 100 galliots,
    in all 160
Leagues of sea coast 2,310
Number of inhabitants in the capital 900,000
Number of cultivated acres, about 128,000,000
Exports to England on an average of 10 years .. 260,000
Imports from England, ditto 280,000
Great divisions of the country, Europe, Asia, Africa, .. 3
Smaller divisions, besides the Greek islands, provinces .. 22
Chief towns, Constantinople 900,000, Aleppo 290,000,
    Cairo 400,000, Ancona 104,000, Smyrna 120,000,
    Adrianople 80,000.
Longitude of central point 37° 15' east.
Latitude of ditto 36°.
Longitude of capital 28° 56' 15" east.
Latitude of ditto 41° 1' north.
Amount of taxes on each person 6s. 10d.

The productions of the Turkish Empire are numerous. Corn of all sorts; great variety of fine fruits. Silk, cotton, coffee, sugar cane, tobacco; copper, and other metals; marble, gum, spices of different sorts; cattle of all sorts; also camels, lions, &c.

THE KINGDOM OF SWEDEN.

SINCE Olow Skanthonung firſt aſſumed the title of King of Sweden, and introduced Chriſtianity there in the year 1000, the revolutions in that kingdom have been numerous. The reigns of Guſtavus Adolphus, the greateſt warrior of his age, and of Charles XII. conferred a temporary ſplendour upon Sweden, which, whilſt neighbouring kingdoms have been riſing and falling, has, amidſt all its own revolutions, and of thoſe around, maintained a very reſpectable rank as a ſecond-rate power.

The royal authority was abſolute till 1718, but from that time the ſtates of the realm gained upon the royal prerogative rather to the detriment of the public weal, until a revolution, very artfully and ably conducted by the late king Guſtavus III. took place in 1792, and the monarchy again became abſolute. Guſtavus was one of thoſe kings who uſed his power to make his ſubjects happy ; neverthelefs he was affaſſinated in 1792, an event regretted univerſally at the time. His brother, the Duke of Sudermania, was regent during the minority of the preſent king, Guſtavus Adolphus IV. who ſhews the ſame diſpoſition with his father, and bids fair to make his ſubjects happy.

The ſucceſſion is hereditary both in the male and female line. Sweden is well ſituated for manufactures and commerce, but neither the one nor the other have been puſhed or encouraged ſo as they might have been. There, as well as in other northern nations, a different ſyſtem is neceſſary for the encouragement of the arts and manufactures from what will ſucceed in warmer climates, and upon a more fertile ſoil.

Extent in square miles 209,000
Number of inhabitants 3,000,000
Number of persons to a square mile 14
Extent in English acres 133,000,000
Number of acres to each person . . . . 44 $\frac{1}{3}$
Number of acres in cultivation 24,000,000
Revenues in pounds sterling 1,500,000
Amount of public debt 7,000,000
Land forces in time of peace 50,000
Ditto, in time of war 140,000
Seamen in time of peace 15,000
Ditto in time of war 35,000
Ships of the line 30
Frigates, sloops, &c.—10 frigates & 60 gallies, in all 70
Extent of sea coast in leagues 380
Number of inhabitants in the capital 80,000
Amount of exports to England 290,000
Amount of imports from England 170,000
Great divisions of the country, Sweden, Gothland,
    Northland, Lapland, Finland, Pomerania, Wismer 7
Smaller divisions, provinces or districts 44
Chief towns, Stockholm 80,000, Gothenburg 20,000,
    Carlscrona 11,000, Stralsund 10,000.
Longitude of capital 18° 3' east.
Latitude of ditto 59° 20' north.
Amount of taxes on each person 10s.
Religion, Lutheran. Calvinist alone tolerated.

  Sweden produces corn, hemp, flax, and cattle of most
sorts. But its main objects of exportation are iron, copper, and timber; hides, skins, and tallow.

THE GERMAN EMPIRE
BEFORE THE WAR.

The principalities of Germany in the 8th century, then united to France, became formidable under the Emperor Charlemagne. In 887 it was separated, and became an independent and diftinct Empire under hereditary princes, but in 1085 became elective, and has ever fince remained fo.

It would be difficult to conceive a more inefficient form of government for a country than fuch a number of princes, all of them entitled to vote in cafe of war, but at perfect liberty to contribute or withhold their contingent in money and in troops.

It is very fortunate that the princes of the houfe of Auftria, which is in itfelf powerful and poffeffed of great territories and revenue, are eligible to the imperial throne, and have been elected; otherwife the Empire would be now the moft fragile political combination that has perhaps ever exifted.

The princes have too many rights to be compelled to co-operation in an effectual manner, from doing which their different views and interefts prevent them. Of this we were lately the witneffes, and are about to contemplate the confequences, when, the Empire being diminifhed in its extent, thofe who have not fuffered owing to local fituation are to be compelled to indemnify others that have.

The German conftitution, of great antiquity, and as it were a middle ftep between the feudal fyftem and limited monarchy, cannot be expected to refift the violent and fyftematical attacks that are in thefe times directed againft every old and eftablifhed form of government.

Extent in square miles 197,000
Number of inhabitants 25,000,000
Number of persons to a square mile 128
Extent in English acres 126,000,000
Number of acres to each person 5
Number of acres in cultivation 90,000,000
Revenues in pounds sterling 14,000,000
Land forces in time of peace 120,000
Ditto in time of war 260,000
Number of inhabitants in the capital 254,000
Amount of exports to England 950,000
Amount of imports from England 1,420,000
Great divisions of the country, 6 Electorates,⎫
  16 Principalities, 11 Ecclesiastical States, ⎪
Lesser ditto, 4 Imperial free cities, and Impe- ⎬ 41
  rial, Prussian, Swedish, and Danish terri- ⎪
  tories, ⎭
Chief towns, Vienna, Berlin, Hamburgh, Liege,
  Munich, Franckfort.
Longitude of central point 12° east.
Latitude of ditto 50° north.
Longitude of the capital 16° 22' east.
Latitude of ditto 48° 12' north.
Amount of taxes on each person 11s. 2d.
Rate of interest of money,
Extent of sea coast, none.
Religion, Roman Catholic, Lutheran, and Calvinist,
  upon a footing of equality.

    The productions are abundant and various. All sorts
of grain, cattle, and fruits; quicksilver, copper, and
other metals. Copperas, allum, tobacco, silk, timber,
olive oil, &c. &c.

THE EMPEROR'S DOMINIONS
BEFORE THE WAR.

WHEN Charles V. who was Emperor of Germany
as well as King of Spain, refigned his imperial and royal
honours and power for a cell in a convent, he left his
German dominions to his brother, and Spain to his fon,
Phillip II.

The male line of Auftria became extinct by the death
of Charles VI. who was fucceeded in his hereditary dominions by his daughter Maria Terefa, married to Francis
Grand Duke of Tufcany, in the year 1740. To Maria
Terefa fucceeded her fon Jofeph II. who was elected
Emperor in 1765. By Galicia, Lodomiria, Buckowena,
and the quarter of the Inn, he added three millions to the
number of his fubjects; and after many well-intended,
but rather unfuccefsful attempts, to make philofophical
reforms among his fubjects, he died difappointed in 1790.
Leopold fucceeded, and reigning only two years, his fon
Francis II. was chofen Emperor.

There is a vast variety of foil in the Emperor's hereditary dominions. The Auftrian Netherlands, and the
Duchies of Milan and Mantua, being remarkably fertile
and well cultivated; Lodomiria and Galicia, taken from
Poland, are likewife very fine countries; and upon the
whole, the Emperor's eftates are much above par with
refpect to fertility and riches.

As the German Empire and the hereditary dominions
are in part the fame, and in part not, it is difficult to
make a clear diftinction betwixt them; one obfervation
may however be made relative to both, which is, that
if ever the ftates of the Empire fhould act in contradic-
tion to the houfe of Auftria, alone more powerful than
all of them together, they will lofe their importance in
Europe, and lay a foundation for their own deftruction.

Extent

Extent in square miles 180,000
Number of inhabitants 19,000,000
Number of persons to a square mile 108
Extent in English acres 115,000,000
Number of acres of land to one person 6
Revenues in pounds sterling 11,000,000
Amount of public debt 40,000,000
Land forces in time of peace 365,000
Ditto in time of war 450,000
Leagues of sea coast (18) 18
Number of inhabitants in the capital 254,000
Number of cultivated acres 75,000,000
Exports to England   } Flanders only ... {   300,000
Imports from England }   { 1,400,000
Great divisions of the country }
Smaller divisions   } 10
Chief towns, Vienna 254,000, Milan 130,000, Brussels 80,000, Prague 80,000, Ghent 60,000, Antwerp 50,000.
Longitude of central point 14° 20′ east.
Latitude of ditto 47° 30′ north.
Longitude of capital 16° 22′ 30″ east.
Latitude of ditto 48° 12′ 32″.
Amount of taxes on each person, 12s.
Religion, Roman Catholic ; but general toleration.

The productions are various. Corn, flax, hemp; cattle, wine, copper, quicksilver, zinc, and other metals. Coal, porcelain, and most sorts of fruit.

THE KINGDOM OF DENMARK.

DENMARK was a few centuries ago one of the moſt warlike nations of Europe, and the people are ſtill of a very brave nature. In addition to their acquiſitions in England, Scotland, and Ireland, which were but of a temporary duration, the Danes ſecured to themſelves the poſſeſſion of Greenland in the 11th century, and of Iceland in the 13th.

Neither the population nor the revenues of Denmark are ſufficient to ſupport it in the rank it formerly held ; it is therefore a ſecond-rate power, and has wiſely contrived for a long period to keep free of quarrels with other nations.

The government is abſolute hereditary monarchy ſince the year 1660, when the people in a voluntary manner made a ſacrifice of their liberties to their king ; from which time Denmark has been in a more flouriſhing ſtate than before. This is a ſtrange fact, contraſted with what during the ſame period has happened to the unfortunate Poles, and is ſufficient to make people ſceptical with regard to all theories about modes or forms of government. During the preſent war Denmark has carried on a great trade, and become much more wealthy than it ever was before ; and no nation in Europe has been ſo free from that political influenza that has prevailed extenſively within theſe laſt twelve years.

The laws of Denmark are all contained in one volume, and juſtice is adminiſtered properly, and at ſmall expence ; which is much more important to the happineſs of the people than any reform that could be effected in the government.

Extent in fquare miles 170,000
Number of inhabitants 2,150,000
Number of perfons to a fquare mile 12
Extent in Englifh acres 108,000,000
Number of acres to one perfon 54
Revenues in pounds fterling 1,520,000
Public debts 2,600,000
Land forces in time of peace 75,000
Ditto in time of war ....
Number of feamen in time of peace 18,000
Ditto in time of war ....
Ships of the line 26
Frigates, floops, &c.—7 frigates of 50 guns, and fmall
 veffels.
Leagues of fea coaft 573
Number of inhabitants in the capital 90,000
Number of cultivated acres 12,600,000
Exports to England 110,300
Imports from England 219,000
Great divifions of the country 3
Smaller ditto 12
Chief towns, Copenhagen, Altona, Elfineur.
Longitude of central point of Denmark Proper 10° 15'.
Latitude of ditto 55° 30' north.
Longitude of chief city 12° eaft.
Latitude of ditto 55° 41' north.
Religion, Lutheran ; others tolerated.
Amount of taxes on each perfon 15s. 3d.

The principal productions are corn, hemp, flax and
cattle. An inferior quality of fir timber is exported from
Norway in confiderable quantity ; but none of the Danifh
dominions are famous for manufactures ; and in fuch a
latitude the finer productions of the earth are not to be
expected.

THE KINGDOM OF SPAIN.

THE foundation of the prefent Spanish monarchy was laid fo lately as the year 1491, by Ferdinand I. who drove the Arabs out of Grenada, after having united Arragon with Caftile, by efpoufing Ifabella, heirefs of the latter kingdom. Previoufly to the time of Ferdinand, Spain had been perpetually over-run by the Arabs, and only dated its greatnefs from that period. For rather more than a century it was the richeft and moft powerful kingdom in Europe ; and it is an opinion entertained, not only by mankind in general but by many of thofe who ought to be better informed, that Spain owes its depopulation and decline to the expulfion of the Moors in 1508, when, in reality, the glory of Spain commenced with Ferdinand, and its decline more than fifty years afterwards, when Charles V. by the conqueft of Mexico and Peru, had opened a door for the influx of gold and the emigration of men. Gold came in by millions every year, and men went over to America in multitudes, with the hopes of fharing in the wealth of newly-difcovered mines. Thefe caufes, together with an ill-conducted government, the neglect of agriculture and induftry, occafioned by the influx of the precious metals, which introduced luxury and idlenefs, gradually reduced the power and importance of Spain, which reduction, the wild projects of Philip II. tended greatly to precipitate.

Spain, once the firft in wealth and power amongft nations, is reduced to a fecondary clafs, and fince the French revolution has fhewn a want of energy that even the poffeffion of unearned gold can fcarcely account for.

The form of government is monarchical and hereditary. The Cortez is a deliberative body intended to controul the executive power, but, like the ftates-general in France before the revolution, has not been called together for many years. When they are, perhaps the confequences will be fimilar.

Extent

Extent in fquare miles 148,000
Number of inhabitants 11,000,000
Number of perfons to a fquare mile 74
Extent in Englifh acres 94,000,000
Number of acres of land to each perfon 8½
Revenues in pounds fterling 14,000,000
Amount of public debt 48,000,000
Land forces in time of peace 104,000
Ditto in time of war 250,000
Number of feamen in time of peace 40,000
Ditto in time of war 104,000
Ships of the line 74
Frigates, floops, &c. 56
Leagues of feacoaft 466
Number of inhabitants in the capital 140,000
Number of cultivated acres 40,000,000
Number of parifhes 19,600
Exports to England on an average laft ten years 600,000
Imports from England ditto, ditto 1,400,000
Revenues of the clergy, of whom there are above 300,000,
    not known.
Great divifions of the country 15
Chief towns, Madrid, Cadiz, Valencia, and Seville.
Longitude of central point 4° 11′ weft.
Latitude of ditto 39° 50′ north.
Longitude of capital city 3° 25′ 15″ weft.
Latitude of ditto 40° 26′ north.
Amount of taxes on each perfon 1l. 5s. 5d.

Spain produces wine, fruits of all forts, olives, corn,
rice, faffron, barilla, and faltpetre. Cattle of all forts;
gold, filver, iron, lead, copper, quickfilver, cinnabar, an-
timony, &c.

BRITAIN and IRELAND.

GREAT BRITAIN was called Britannia by the Romans, who invaded it fifty-five years before the birth of Chrift, from which time, till the year 446, it remained under their yoke. The Danes and Saxons ruled alternately till the invafion of William the Conqueror in 1066. In 1172 Ireland was conquered, and in 1284 Wales. In 1603 the crowns of England and Scotland were united under James I. in 1706 their parliaments were united, and in 1800 the English and Irish parliaments, fo that there is now only one parliament for the three kingdoms.

The form of government is monarchical, the fuc-ceffion to which is hereditary in both lines in the houfe of Brunfwick. The legiflative power vefted in king, lords, and commons, but the executive in the king affifted by a council of his own nomination*.

England is now the firft commercial and manufactur-ing nation; it is alfo the greateft naval power. Its re-venues and expenditure are beyond thofe of any other nation.

The ufe of machinery has been carried to an immenfe length, and its conftruction to great perfection, fo that the labour of more than three millions of perfons is per-formed by inanimate workmen as they may be termed, who both toil and fpin without requiring either food or raiment, the keeping of which in repair, added to the intereft of the firft expence, does not amount to above three halfpence a day on the labour of one perfon worth a fhilling, the agregate gain on which is three millions of French livres in one day, or £.126,000!! It is owing chiefly to fuch inventions that this nation is able to fup-port its great debts and annual expences.

England is fruitful and well cultivated, but exports little of its produce. Of late years the corn produced has not been equal to fupply the country.

---

* This form of government is the beft yet eftablifhed in any country, being a happy mean between abfolute monarchy and the turbulent fyftems of federations and perfect equality.

Extent

Extent in square miles 104,000
Number of inhabitants 14,000,000
Number of persons to a square mile 136
Extent in English acres 67,000,000
Number of acres to each person 4¾
Number of acres in cultivation 40,000,000
Revenues in pounds sterling.............. 27,000,000
Amount of public debt, ... 400,000,000
Land forces in time of peace 45,000
Ditto regulars and militia of all sorts this war .. 350,000
Seamen in time of peace 18,000
Ditto in time of war 112,000
Ships of the line 187
Frigates, sloops, &c. 441
Extent of seacoast in leagues 1,200
Tonnage of merchant ships 1,800,000
Number of inhabitants in the capital 1,100,000
Number of parishes, 9,000 in England and 1000 in
    Scotland (not including Ireland) 10,000
Exports to all parts, average 30,000,000
Imports from all parts, ditto 25,000,000
Expence of maintaining the poor.......... 3,000,000
Expence of the clergy 7,000,000
Great divisions of the country, England, Scotland, Wales,
    Ireland 4
Smaller divisions, counties 117
Chief towns, London, Dublin, Edinburgh, York, Liver-
    pool, Bristol, Newcastle.
Longitude of central point 1° 3′ west.
Latitude of ditto 53° 40′ north.
Longitude of the capital city 0° 0′, this and most English
    books calculate from the meridian of London.
Latitude of ditto 51° 31′ north.
Amount of taxes on each person 1l. 18s. 3d.
Rate of interest of money, 5 per cent. in England and
    Scotland, 6 per cent. in Ireland.
Religion, Protestant, Lutheran and Calvinist ; all sects
    tolerated.

C 3

THE KINGDOM OF PRUSSIA.

So late as the year 1656 Pruffia was only a fief of the kingdom of Poland, of which it is now one of the mafters. It was rendered independent of Poland by C. Frederick William, then Duke of Pruffia and Marquis of Brandenburgh, but a warlike and great prince.

Pruffia firft rofe to the rank of a kingdom in 1701, under Charles Frederick III. whofe political conduct was fuch as to make the fmall dominions over which he ruled of fo much importance, that his title was acknowledged by all the powers of Europe.

It was when Frederick II. better known by the name of Frederick the Great, afcended the throne in 1740, that Pruffia began to be confidered as one of the leading powers in Europe, to which title, that great monarch, before he died in 1786, fully eftablifhed its claim. He gradually increafed the extent of his dominions, mantained defperate and expenfive wars againft formidable neighbours, yet terminated them with advantage and glory. Notwithftanding his wars, and with a very limited revenue, Frederick expended annually more than half a million fterling in the encouragement of arts, and in advancing internal profperity ; and while the great and wealthy nations of Europe were running in debt, this abfolute monarch died with a full treafury, leaving as his laft charge to his high chancellor, an order to draw up a better code of laws for the fubjects of his fucceffor.

Extent

Extent in fquare miles 56,000
Number of inhabitants 5,500,000
Number of perfons to a fquare mile 90
Extent in Englifh acres 34,000,000
Number of acres to each perfon 6
Revenues in pounds fterling 4,200,000
Land forces in time of peace 224,000
Ditto, in time of war 350,000
Leagues of feacoaft 50
Number of inhabitants in the capital 80,000
Number of cultivated acres 25,000,000
Great divifions of the country 6
Smaller ditto 27
Chief towns, Berlin, Breflaw, Konigfberg.
Longitude of central point
Latitude of ditto
Longitude of capital 13° 22′ 30″ eaft.
Latitude of ditto 52° 31′ 30″ north.
Amount of taxes on each perfon 14s. 6d.
Eftablifhed religion, Proteftant.

The productions of Pruffia are corn of all forts. Fruits, flax, hemp, hops, horfes, cattle, fheep, timber, metals, &c.

NAPLES AND TWO SICILIES.

LIKE other Italian ſtates, Naples and the Sicilies have undergone numerous changes which have generally been of little importance except for the moment. After having been alternately in the hands of the Germans, French, and Spaniards, for ſeveral centuries, Ferdinand IV. the third ſon of Charles III. king of Spain, was created king of the Sicilies in 1754, and commenced his reign in 1767 with an expreſs ſtipulation that Naples and the Sicilies ſhould never again be united to the crown of Spain.

The Neapolitan dominions are by nature fruitful and rich beyond almoſt any other country; but as the people are idle, turbulent, and mutinous, Naples never has either enjoyed power or tranquillity. There is a claſs of people here, unknown in any other European nation, and diſtinguiſhed by the name of Lazaroni, who by the favour of ſo fine a climate are enabled to live almoſt altogether in the open air, and by a ſpecies of diſcipline amongſt themſelves, and their great numbers, they are formidable both to the court and the people. What changes may reſult from the preſent war it is not eaſy yet to ſay; but the beſt guarantee ſeems to ariſe from the family connection with the thrones of Spain and Auſtria.

Extent

Extent in square miles 30,000
Number of inhabitants 6,000,000
Number of persons to a square mile 200
Extent in English acres 19,200,000
Number of acres to each person 3⅓
Revenues in pounds sterling 1,400,000
Land forces in time of peace 34,000
Ditto in time of war 80,000
Seamen in time of peace 5,500
Ditto in time of war 8,000
Ships of the line 4
Frigates, sloops, &c. galliots, gallies, &c. 23
Extent of seacoast in leagues 586
Number of inhabitants in the capital 380,000
Revenues of clergy, estimated at one-half the revenues,
    and one-third of the lands in the kingdom, 200,000 in
    number.
Great divisions of the country 4
Smaller divisions 19
Chief towns, Naples, Palermo, Bari, Catanea.
Longitude of central point in Italy 15° 10′ east.
Latitude of ditto 41° north.
Longitude of capital 14° 12′ east.
Latitude of ditto 40° 50′ north.
Amount of taxes on each person 4s. 9d.
Religion, Roman Catholic.

Naples and the Sicilies produce corn, excellent fruits,
olives, wine, rice, tobacco, cotton; cattle of all sorts,
gold, silver, iron, marble, alabaster, pit coal, &c.

THE KINGDOM OF PORTUGAL.

PORTUGAL may be confidered as Spain in miniature, being in fituation, foil, and climate, nearly fimilar. Like Spain it flourifhed and was wealthy on account of its poffeffions abroad, and like Spain it has funk from the importance it once enjoyed.

The form of government is defpotic, the fucceffion hereditary in either fex in the houfe of Braganza. The Portugeze were the firft that doubled the Cape of Good Hope, as well as that difcovered the Brazils, in the end of the 16th century, and for a confiderable period were, next to Spain, the moft brilliant and wealthy people in Europe; but, like Spain, Portugal is a monument of the evanefcent nature of wealth arifing from foreign poffeffions. Agriculture, induftry, and manufactures, which keep up the good habits of a people, are true and permanent fources of profperity; but an influx of gold deftroys thofe true fources, and replaces them with falfe ones, which, gradually difappearing, leave a nation in liftlefs inactivity, incapable of even maintaining the rank to which it is naturally entitled.

The precious metals which it imports from the Brazils remain but a very little time in Portugal, being employed to buy manufactured goods from other more induftrious nations. Thefe are to the amount of about two millions fterling per annum.

Extent

Extent in fquare miles 27,000
Number of inhabitants 1,838,000
Number of perfons to a fquare mile 67
Extent in Englifh acres 17,280,000
Number of acres of land to one perfon 10
Revenues in pounds fterling 2,150,000
Amount of public debt 4,000,000
Land forces in time of peace 36,000
Ditto in time of war 60,000
Seamen in time of peace 12,000
Ditto, in time of war 22,000
Ships from 40 to 80 guns 18
Frigates, floops, &c..........................40
Number of inhabitants in the capital.........120,000
Number of parifhes (and above 200,000 eccle-
    fiaftics) 3,500
Great divifions of the country 5
Amount of taxes on each perfon, 1l. 3s. 2d.
Extent of feacoaft in leagues 166
Chief towns, Lifbon, Oporto.
Longitude of capital (moft wefterly town in Europe)
    9° 9' 15" weft.
Latitude of ditto 38° 42' 20" north.
Longitude of central point 8° 20' weft.
Latitude of ditto 39° 30' north.
Religion, Roman Catholic ; they are not tolerant to
    other religions.

The productions of Portugal are the fame with thofe
of Spain. The particular fpecies of wine called port
is in great requeft towards the north of Europe,
and in England more than any other country. The
quantities of this wine that are produced are very great,
and make the principal article of exportation from Portugal.

SARDINIA AND SAVOY.

THIS kingdom confifts of the ifland of Sardinia in the Mediterranean fea, and the dutchy of Savoy on the north-weft of Italy, together with the country of Piedmont, with fome other dependencies.

It is one of thofe kingdoms which has owed its political importance chiefly to the talents and family connections of the reigning princes.

Strongly fituated amongft the Alps, with a vigorous and uncorrupted race of inhabitants, and a line of princes equally brave and virtuous, the continental dominions, though fmall, fupported a refpectable ftate of independence, and their princes, though never chief in any war, were confidered as defirable allies or dangerous enemies by thofe who did engage in military contefts. Since the year 1016 the prefent race have governed in Savoy, although it is only fo late as the year 1718 that Sardinia was added, and the title of kingdom conferred on thofe united poffeffions.

Now that war has become fo expenfive, the importance of fmall ftates with little revenue muft decreafe rapidly; and fuch is the cafe with the kingdom now under confideration.

Extent

Extent in fquare miles 20,000
Number of inhabitants 3,253,000
Number of perfons to a fquare mile 162
Extent in Englifh acres 12,800,000
Number of acres of land to one perfon 4
Revenues in pounds fterling 1,850,000
Amount of public debt, none.
Land forces in time of peace 38,000
Ditto, in time of war 100,000
Seamen in time of peace 6,000
Ditto, in time of war 10,000
Ships of the line, frigates, floops, galliots, galleys, &c. } 32
   veffels of all forts armed ....
Extent of feacoaft in leagues ....
Number of inhabitants in the capital 82,000
Great divifions of the country 5
Smaller divifions 19
Chief towns, Turin, Vercelli, Cagliari.
Longitude of central point, continental dominion, 7° 30′
   eaft.
Latitude of ditto 45° north.
Longitude of capital 7° 40′ eaft.
Latitude of ditto 44° 5′ north.
Amount of taxes on each perfon 10s. 6d.
Religion, Roman Catholic, but tolerant.

Savoy is rather a barren country, but Piedmont and Sardinia abound in the productions of Italy, corn, wine, oil,
fruits of all forts, and great numbers of cattle; filk is alfo
produced in very confiderable quantities.

SEVEN UNITED PROVINCES.

THE whole of the 17 provinces, which belonged to the Dukes of Burgundy, devolved to the houfe of Auftria in 1477 by marriage, and afterwards by marriage alfo to the crown of Spain; but a number of thofe provinces foon began to ftruggle for liberty, and after an uncommon difplay of bravery and perfeverance during the long term of 80 years, feven of them obtained that independence which they had fo well deferved. Holland being the chief of thefe feven provinces, it has been cuftomary to call the whole union by that name.

Holland became the greateft commercial country in the world, confequently a very rich and refpectable power both by fea and land, but more particularly fo by fea.

This profperity, however, as ufual, was not of very long duration; for though it did not bring indolence and luxury into Holland in the fame manner that it had done into Spain and Portugal, yet induftry did relax, and the merchants who ufed to fpeculate for themfelves were contented with receiving the fmall but certain profits of agents for others, from which time the Dutch importance has been on the decline. Difcontent and faction have tended greatly to reduce the country, which, from being a firft-rate power, has now fallen to lefs than a fecond-rate, or rather to that of a fubjected province of France; but this will probably not long continue.

Extent in fquare miles 10,000
Number of inhabitants 2,758,000
Number of perfons to a fquare mile 257
Extent in Englifh acres 6,400,000
Number of acres of land to each perfon .... $2\frac{1}{3}$
Revenues in pounds fterling 3,500,000
Public debts in ditto 11,000,000
Land forces in time of peace 36,000
Ditto in time of war ....
Seamen in time of peace 16,000
Ditto in time of war 40,000
Ships of the line 40
Frigates, floops, galliots, &c. 50
Number of inhabitants in the capital 212,000
Number of parifhes about 1,600
Amount of exports to England 600,000
Amount of imports from ditto 1,900,000
Leagues of feacoaft 236
Great divifions of the country 9
Chief towns, Amfterdam, Rotterdam, Leyden, Harlem,
    the Hague
Longitude of capital 5° 4′ eaft.
Latitude of ditto 52° 22′ north.
Longitude of central point 5° 30′ eaft.
Latitude of central point 54° north.
Amount of taxes on each perfon 1l. 12s. 3d.
Religion, Calvinift ; but tolerant to all others.

No country is better cultivated or more productive for
its extent; but the population is fo great that moftly all is
confumed in the country. Butter, cheefe, and falted provifions are however exported, and fifhing is followed
with great induftry and fuccefs by that indefatigable
people. Every fpecies of induftry is on the decline.

Befides

*Observations on, and explanation of Chart 2d, representing
the principal nations of Europe, as they stand, according to
the new order of divisions and alliances.*

PREVIOUS to the French revolution Europe was in a
serene and tranquil situation, which may not improperly
be compared to the placid and smooth surface of that
great river in North America, which empties the waters
of the immense superior lakes into the inferior lake On-
tario, before that prodigious mass of water which it contains precipitates itself over the huge rocks of Niagara.
The same mass of water which before moved on serene
and flow, after the sudden and tremendous fall, boils up
and eddies in a thousand directions, changing at every
instant with irregular impetuosity, until distance of space
and length of time again restore to the disturbed element
its natural calm and regular movement.

We have represented the river previous to its fall ; we
are now at the bottom of the cataract, and it remains for
us either to take a view of it in its present turbulent situation, or to desist until the lapse of time and a succession
of events shall again have restored order and tranquillity.

The situation of Europe is too important to let all
pass on unnoticed, until a day, certainly not very near at
hand, and probably at a considerable distance, shall arrive,
when a permanent and solid peace may be established. It
is perhaps not going too far to say, that much utility and
real advantage may arise from representing the state of the
governments of Europe as they will be, supposing the
treaty already entered into between France and Austria
to take place, and be realized in a durable manner.

We mean to say, that a representation made out before
matters be finally settled, may compensate for what it

D   wants

11. Number of clergy and amount of their revenues.
12. Leagues of inland navigation.
13. Number of horses.
14. Criminals executed.
15. Ditto transported.
16. Ditto imprisoned, tried, &c.
17. Ditto acquitted.
18. Current coin in circulation, amount of.
19. Number of banks.
20. Paper circulation, estimate of.
21. Grain exported, } average.
22. Ditto imported, } average.
23. Number of persons imprisoned for debt, average.
24. Average income or expence of each individual.
25. Total quantity of corn consumed.
26. Quantity of work done by machinery.
27. Quantity of power of fire engines, measured by the strength of horses.
28. Price of travelling post with two horses.
29. Number of bankruptcies.

Note. The pale red circle round France shews the extent of that country, together with those under the authority of its present rulers.

STATISTICAL ACCOUNT

OF

HINDOOSTAN.

THIS interefting portion of the globe is comprehended between the 70th and the 90th degrees of eaft longitude, and the 8th and 35th of north latitude. Its general boundaries are, to the north, the kingdom of Thibet, from which it is feparated by the mountains of Hindoo Khoo; on the fouth, by the great Indian ocean; on the eaft, by the Burrampooter river and the bay of Bengal; and on the weft, by the Indus, Perfia, and the Arabian gulph.

The population of Hindooftan is not fo confiderable as might be expected; but it muft be confidered that although Britifh India is extremely populous, there are other countries very thinly inhabited.

The revenues of Hindooftan have, fince the reign of Aurengzebe, who died in 1707, been on the decline. The provinces of Bengal and Bahar have, it is true, under the prudent adminiftration of our late Governors-general of India, experienced a contrary effect. Britifh India by the continuance of the fame falutary meafures under the prefent adminiftration, is daily acquiring an increafe of population and revenue.

The fituation of Hindooftan is admirably fuited for commerce, both inland and maritime. Its extent of fea-coaft gives it almoft all the advantages of an ifland, efpe-cially the peninfula, and the produce of Hindooftan Proper is conveyed to the ports on the gulf of Arabia, and the bay of Bengal, by the Indus, the Ganges, and the Barrampooter.

The inland commerce of Hindooftan is carried on by the means of caravans with Bootan, Thibet, Siam, Tar-

D 3

tary,

company to improve the breed of this useful animal.
There are also black cattle, sheep, elephants, camels,
ravenous animals, such as tigers, wolves, bears, &c.
Deers and antelopes in a great variety, wild hogs, hares,
partridges, snipes, wild ducks, and all sorts of domestic
fowls.

The manufactures of Hindoostan are chiefly those of
cotton and silk; from the first they derive the most beautiful muslins in the world, with the greatest variety of
cotton cloths of all descriptions. They also manufacture
saltpetre, rum, sugar, arrack, indigo, and salt. The natives
work curiously in gold and silver, and they embroider
on the finest muslin, and on cloth, to admiration. They
are good mechanics, and expert ship-builders.

In a country enjoying the benign effects of a salubrious
climate, where little cloathing is necessary, the inhabitants simple in their manners, and whose modes of life
are abstemious in the extreme, are enabled to produce
articles, both of necessity and luxury, at a price so moderate, as to enable those who possess the commerce of
Hindoostan to undersell every market in the world.
The price of labour does not exceed sixpence a day, and
the artizan may possibly earn a third more than that
sum.

Land produces to the state from ninepence half-
penny to one shilling and nine-pence farthing per acre,
whilst the share to the cultivator is less than one third
of the actual produce. It is not so much amongst the
native powers, the governments or the rulers of Hindoostan, as the zemendaurs and their dependents, the
cutwal, or judge, and the collectors of the duties and
customs, who oppress the unfortunate natives of Hindoostan.

STATISTICAL TABLE OF

HINDOOSTAN.

Extent of Hindooftan in fquare miles 1,024,800
Number of inhabitants 77,986,818
Number of perfons to a fquare mile, in different pro-
    vinces 62. 80. 114. 125
Number of Englifh acres 655,872,000
Number of acres to each perfon about 8½
Revenue in pounds fterling 30,000,000
Commercial exports, about 7,000,000
Imports 3,000,000
Extent of fea-coaft in leagues, about.......... 1,200
Peninfula of India in fquare miles 167,911
Extent of the Merhatta empire in fquare miles 457,144
Britifh poffeffions 217,185
Britifh allies 235,467
Britifh interefts in India in fquare miles 452,652
Number of inhabitants in ditto 41,062,890
Revenue of ditto 15,459,000
Nizam's territories 103,690
Revenues in pounds fterling 2,600,000
Military ftrength, cavalry 40,000 infantry 30,000 70,000
Dominions of the late Tippoo Sultan before the partition
    in 1792 in fquare miles 98,000
Revenue........................... 2,380,000
Dominions of Tippoo after the partition in 1792 in
    fquare miles, ... 62,000
Revenue 1,425,000
Divifion of the empire of Myfore in fquare miles, to the
    Britifh about 32,000
To the Raja of Myfore 26,000
To the Merhattas 13,000
To the Nizam 26,000
Revenue of the Mogul empire in the reign of Aurung-
    zebe 32,000,000

Extent

Extent of ditto in square miles 827,415

District of Delhi, the present Mogul empire, about square miles 1,600

Population of the Merhatta empire 28,342,928

Revenue of the Merhatta empire including the Chout 16,000,000

Military strength of ditto, cavalry 210,000, infantry 64,000, total 274,000

Revenues of the Poona Merhattas in pounds sterling 4,000,000

Extent of territory in square miles 152,381

Military strength 60,000

Revenues of Dowlut Row Scindeah in pounds sterling 6,000,000

Military strength, 60,000 cavalry, 30,000 infantry 90,000

Revenues of the Bouncila in pounds sterling 3,500,000

Military strength of ditto, 50,000 cavalry, 10,000 infantry 60,000

Revenues of Holkar in pounds sterling 1,500,000

Military strength of ditto, 30,000 cavalry, 4000 infantry 34,000

Revenues of Guyacquar in pounds sterling .. 1,000,000

Military strength, (cavalry) 30,000

Revenues of the Seicks 1,457,400

Military strength (principally cavalry) 100,000

Extent of the territory of the Seicks in square miles 89,900

Extent of Zemaun Shaw's dominions in square miles 320,000

Population 19,000,000

Revenue 8,000,000

Population of the independent states including the districts of Goa, Cashmere, &c. 1,888,000

Extent of ditto in square miles 23,600

STATISTICAL TABLE OF
BRITISH INDIA.

Actual poffeffions in fquare miles.

Bengal, Bahar, Oriffa, and Benares 162,256
Circars 17,508
Coimbetore .... 10,150 ⎫ Part of
Barramahal .... 7,400 ⎪ the late
Malabar and Coorg .... 6,600 ⎬ kingdom
Canara and part of Soonda .... 6,235 ⎪ of
Dindegul .... 2,600 ⎭ Myfore.
Jaghire in the Carnatic 2,436
Iflands of Bombay and Salfette 2,000
———————217,185

Allies and Tributaries.

Nizam 103,690
Oude 52,880
Carnatic, Tanjore, &c. 44,297
Myfore 25,250
Cochin and Travencore 9,350
———————235,467

Total of the Britifh interefts in India in fquare
  miles 452,652
Total number of inhabitants in ditto 41,062,890
Revenues of ditto, about 19,000,000
Number of inhabitants in Britifh India 23,057,300
Average number of people to a fquare mile 105

Population.

Bengal, Bahar, Oriffa, and Benares 18,497,184
Circars, Coimbetore, Barramahal, and Din-
  degul 2,636,060
Malabar and Coorg 825,000
Canara and part of Soonda 749,066
Jaghire * 170,000
Iflands of Bombay and Salfette 180,000
Nizam 6,428,780
Oude 5,288,800
Carnatic 3,543,760

* In this calculation, the population of the black town of Madras is not
included; neither is that of Seringapatam, now a Britifh garrifon, nor
Madras, included in the total number of inhabitants in Britifh India.

Myfore

Myfore 1,565,500
Cochin and Travencore 1,168,750
Revenue of Britifh India in pounds fterling .. 9,742,937
Charges 8,961,180
Net revenue 781,757
Company's imports from India annually, to the amount
   of, in pounds fterling, about............ 2,000,000
Debt of the company 14,000,000
Intereft of debt paid by the company 978,856
Intereft of money, variable from 6 to 12 per cent.
Extent of Bengal in fquare miles 97,244
Extent of Bahar, Britifh Oriffa, and Benares ... 65,012
Revenue of Bengal, Bahar, Oriffa, & Benares, 6,504,738
Charges 4,332,991
Number of inhabitants in Bengal 11,000,000
Number of perfons to a fquare mile 114
Number of Englifh acres in Bengal 62,236,160
Number of acres to each perfon .... $5\frac{1}{2}$
Revenue of Oude 2,500,000
Revenue of the prefidency of Fort St. George 2,822,536
Charges 3,132,919
Revenue of the Circars 430,000
In 1796 Bengal exported to the value of 3,778,704
Same year imported 1,563,200
In 1796 Fort St. George exported to the value of 802,457
Same year imported 381,568
Revenue of Bombay 415,663
Charges 1,495,270
Number of fquare acres on the iflands of Bombay and
   Salfette 1,280,000
Number of perfons to a fquare mile 90
Number of acres to each perfon 7
In 1796 Bombay imported to the value of 245,537
In the fame year exported 143,925
Extent of territory obtained from Tippoo in 1792 in
   fquare miles including Coorg 16,600
Revenue obtained per annum 395,000
Extent of territory obtained in 1799 in fquare miles 16,385
Revenue obtained * 539,056

* In this is included feven lacs of pagodas, or £.280,000 fterling, ftipu-
lated to be paid by the Rajah of Myfore to the company.

Total

Total territory obtained 32,985
Total revenue ditto 934,056
Extent of seacoast in leagues, about 700
Course of navigable rivers in Bengal in British miles 1,640
Extent of seacoast to ditto in leagues 130
Number of persons to a square mile in Malabar, Cochin,
   and Travancore 125
Total number of inhabitants 2,000,000
Total number of square miles 15,950

Eaſt India Company's Land Forces, including the King's
troops ſerving in India.
Regiments of European cavalry, four 2,400
.... of native ditto, nine 5,400
.... of European infantry, twenty-four .. 24,000
.... of native ditto, forty-two 84,000
Battalions of artillery, ſix 3,000
Corps of engineers, pioneers, &c. 500
Total, independent of irregulars 119,300
Number of Europeans reſiding in India under the pro-
   tection of the Company not in their ſervice .. 1,707
Civil ſervants of the company 2,814
Military officers, including ſurgeons 2,869
Naval officers at Bombay 113

Company's Marine.
Ships 4, ſnows 3, ketches 4, brigs 2, ſchooners 2, beſides
   cutters, packets, &c.
Total number of Britiſh in India, ſubject to the control
   of the Eaſt India Company 35,003

Price of labour in Hindooſtan, equal to one-fourth of the
price of labour in Great Britain, viz.
A common labourer per month of 30 days, calculating
   the rupee at two and ſixpence, 12s.
A perſon who carries burthens 15s.
A bricklayer 18s. 9d.
A maſon 18s. 9d.
A Blackſmith 22s. 6d.
A carpenter 22s. 6d.
A native ſoldier's pay 20s.

HAVING now reprefented, with as much accuracy as the proofs and documents yet collected will admit, the ftate of the population, revenues, &c. of different countries, but more particularly thofe of Britain and Britifh India, a few obfervations on the latter may not be confidered as improper.

Our Britifh poffeffions in India, unlike any other foreign territory belonging to us, are not directly fubject to the government of this country, but are regulated as it were at fecond-hand, by the intervention of the Court of Directors, who are controuled by a Board of Commiffion for regulating the affairs of India, and in fome inftances fubject alfo to the revifion of a General Court of Proprietors. Thus fettered the Directors difpatch their orders for the government of a country at a diftance of eight thoufand miles, of which the extent and population are double thofe of Great Britain, and producing more *free revenue* than the Britifh government poffeffes after the intereft of its debt is paid *.

This fubject is very intricate, and has of late occupied the minds of many able men. To enter into details here would be abfurd ; but we may take a view of the refult.

India cofts this nation a great deal, and has been the caufe of much envy towards this country, the burthens on which have become enormous; not by lavifh expenditure in time of peace, but by the expences occafioned by repeated wars : and it would appear fair, that while the mother country dedicates three fourths of its revenue to the payment of intereft, India fhould contribute fomething ; and that the expences of the eftablifhments there fhould not be allowed to keep pace with, and abforb nearly the whole of the revenues collected.

---

* The free revenue of Britain does not amount to feven millions after the intereft of its debt is paid. That of the Indian territory paffes eight millions after the intereft of fourteen millions is difcharged.

The

The wages of labour are not indeed an exact criterion by which the value of money may be estimated; but all writers on political economy and finance allow them to be one of the best; and as wages are only about one-fourth in that country of what they are in this, it follows, that nine millions there muſt be a very enormous revenue. It is true that a number of individuals are, and muſt be, largely paid there; but in that, as in every government on the face of the earth, the far greater portion of the expenditure goes for the payment of ſubordinate perſons, ſuch as ſoldiers, and thoſe whoſe pay is proportioned to the expence of their exiſtence, the maintenance of horſes, purchaſe of ſtores, &c.

The princes of the country maintain ſplendid courts, yet they amaſs wealth; but without any ſuch royal ſtate to maintain, the company have great debts and no treaſure. Such is the actual reſult, concerning the cauſes of which it would be well deſerving the attention of thoſe who are in power to inquire.

The commerce with the Eaſt, which is likewiſe the envy of all nations, and which, from the earlieſt period, has brought enemies upon every country that poſſeſſed it, is at preſent under a ſtrange predicament. Our India Company appear to monopolize the whole of it; but in reality, ſuch laws have been made to protect the company, that four fifths of it is eſtimated as baniſhed, and in the hands of ſtrangers, ſo that we who ſeem to engroſs all, have in fact only a very inferior portion.

On this important ſubject, however, there exiſt opinions in their nature diametrically oppoſite. By one party it is maintained, that any abbreviation of the company's excluſive charter would endanger its exiſtence; while the advocates for a free trade, with equal confidence aſſert, that not only its welfare as a corporate body, but the proſperity of Britiſh India, the public revenue, and commercial intereſts of this country, would, by a fair participation, be greatly augmented. Certain it is, that, until the expiration of the charter, no arrangement, without the conſent of the company themſelves, can poſſibly be formed.

As

As fome eſtabliſhment, fimilar to that of the preſent
company, muſt always be neceſſary to conduct the affairs
of India, to prevent the deleterious effect of unbounded
patronage, it is fincerely to be wiſhed, that the Court of
Directors and the Board of Controul could devife fome
conciliatory mode by which that part of the commerce
which they cannot embrace may be conceded. At preſent,
the furplus trade of India finds its way into other countries, from the Britiſh merchants being in a manner ex-
cluded from fending home, in ſhips built in India, the
valuable produce and manufactures of that country.

On the 10th of Auguſt 1801, will be publiſhed,

By J. WALLIS, Paternoſter-Row,

THE COMMERCIAL AND POLITICAL ATLAS,

SHEWING THE

Trade and Revenues of Great Britain for the whole of
the laſt Century,

ILLUSTRATED BY STAINED COPPER PLATE CHARTS.

By WILLIAM PLAYFAIR.

This work is printed of a fize to bind up with the preſent,
and both will be found an agreeable and neceſſary
companion for an academy or counting houſe.

ADVERTISEMENT.

In the course of conftructing thofe charts, it occurred that the beft mode of making a ftatiftical and agricultural furvey of England, would be to take each county feparately by itfelf, and reprefent the eftates of all the proprietors who poffefs more than one hundred acres of land, by a fquare of a proportional fize, following each other in the order of their extent. The cultivated lands, foreft lands, and wafte lands, would be diftinguifhed by a difference in the colouring. The name of the proprietor, number of houfes, perfons, cattle, &c. would be marked on each eftate of more than 500 fquare acres; the contents of a chart would be as under, with refpect to manner :

SUPPOSED STATISTICAL ACCOUNT OF THE COUNTY

OF BEDFORD.

Total extent of the county, 617,000 fquare acres.

2 Eftates above 20,000 acres each 52,000
3 Ditto above 10,000 and below 20,000 .. 48,000
7 Ditto above 5,000 and under 10,000 50,000
10 Ditto above 4,000 and under 5,000 43,000
26 Ditto above 3,000 and under 4,000 80,000
50 Ditto above 2,000 and under 3,000 60,000
150 Ditto above 1,000 and under 2,000 100,000
100 Ditto above 500 and under 1,000 36,000
250 Ditto above 100 and under 500 20,000
Wafte lands, roads, &c. ,.................50,000

617,000

With appropriate explanation, care, and accuracy, a true ftatiftical account of England might in this manner be obtained, and that at no very great expence. The author has at this time an intention of publifhing a propofal for this purpofe, and for one county only ; in which cafe a fubfcription will be neceffary, and that lodged in the hands of a banking houfe till the delivery of the work.

LONDON, 30th *July* 1801.

T. Bensley, Printer, Bolt Court, Fleet Street.